中日韩三国

环境科学研究院（所）长
会议机制发展回顾与展望

Review and Outlook of Tripartite Cooperation among
National Environmental Research Institutes of

China, Japan, and
the Republic of Korea

主　编：张孟衡　吴婧赟

副主编：周羽化　加依娜尔·波拉提

中国环境出版集团·北京

图书在版编目（CIP）数据

中日韩三国环境科学研究院（所）长会议机制发展回
顾与展望：中英对照 / 张孟衡，吴婧赟主编. — 北京：
中国环境出版集团，2025.1
ISBN 978-7-5111-5635-8

Ⅰ.①中… Ⅱ.①张… ②吴… Ⅲ.①环境科学－研
究院－会议－制度－研究－中国、日本、韩国－汉、英
Ⅳ.①X-24

中国国家版本馆CIP数据核字（2023）第193353号

责任编辑	邵　葵	
封面设计	宋　瑞	

出版发行	**中国环境出版集团**	
	（100062　北京市东城区广渠门内大街16号）	
	网　　　址：http://www.cesp.com.cn	
	电子邮箱：bjgl@cesp.com.cn	
	联系电话：010-67112765（编辑管理部）	
	发行热线：010-67125803，010-67113405（传真）	
印　　刷	北京中科印刷有限公司	
经　　销	各地新华书店	
版　　次	2025 年 1 月第 1 版	
印　　次	2025 年 1 月第 1 次印刷	
开　　本	787×1092　1/16	
印　　张	11.25	
字　　数	184千字	
定　　价	90.00元	

中国环境出版集团郑重承诺：
中国环境出版集团合作的印刷单位、材料单位均具有中国环境标志产品认证。

本书编委会

主　　编：张孟衡　吴婕赟

副　主　编：周羽化　加依娜尔·波拉提

编委会成员：马会然　闫丽娜　刘可欣　吴　倩

前　言

PREFACE

20世纪80年代以来，东北亚地区工业化和产业化快速推进，经济高速增长，在创造社会财富的同时，也面临着环境和资源的"双重"制约，区域性环境压力凸显。进入21世纪，环境作为可持续发展三大支柱之一，受到各国的关注和重视。作为区域和全球的重要经济体，中国、日本、韩国（以下简称中日韩）认识到区域和全球环境问题的解决已超出了单个国家的能力，需要各国在区域和全球范围内加强环境科技交流与合作，才能有效解决复杂的环境问题。

中日韩三国环境科学研究院（所）长会议（TPM）机制，于2004年由中国环境科学研究院倡议，与日本国立环境研究所、韩国国立环境科学院共同建立。三方本着"友谊、交流、合作、共赢"的基本原则，通过定期会议机制促进学术交流和相互理解、分享环境治理经验、探索环境解决方案；通过交流研讨确立合作愿景、制订主要研究议程；通过持续合作共同应对新的挑战、推动东北亚地区环境质量持续改善。

在过去的20年间，无论外部环境如何变化，TPM机制下的合作从未中断，合作机制不断得到发展和完善，经历了夯基垒台、稳步拓展、务实深化三个发展阶段，已成为中日韩三国环境科学研究院（所）开展科研交流、互学互鉴、合作共赢的重要平台和有效机制，取得了丰硕成果，对东北亚地区乃至全球可持续发展发挥了积极的作用，为区域环境合作树立了成功典范。

在TPM机制成立20周年之际，谨以此书回顾三方合作发展历程，展现合作成果，总结发展经验，展望合作前景，以期为区域及全球环境科技领域的国际合作交流提供参考。

第一部分
TPM 机制发展回顾与展望　001

1 TPM 机制建立背景、历次会议主要成果及会议举办方式 ……………002

  1.1　TPM 机制建立背景 …………………………………………… 002

  1.2　TPM 历次会议主要成果 …………………………………… 004

  1.3　会议举办方式 ………………………………………………… 009

2 TPM 机制发展历程 ………………………………………………… 016

  2.1　合作起步阶段：TPM1 ～ TPM4（2004—2007 年）………… 016

  2.2　稳步拓展阶段：TPM5 ～ TPM13（2008—2016 年）……… 019

  2.3　务实深化阶段：TPM14 至今（2017 年至今）……………… 022

3 科研领域合作进展 ………………………………………………… 027

  3.1　大气环境领域 ………………………………………………… 027

  3.2　水环境领域 …………………………………………………… 029

  3.3　气候变化领域 ………………………………………………… 031

  3.4　环境健康领域 ………………………………………………… 033

  3.5　其他科研领域 ………………………………………………… 035

  3.6　小结 …………………………………………………………… 038

4 科研体制机制交流成果 …………………………………………… 041

5 面临的挑战与未来合作展望 ……………………………………… 045

6 结语 ………………………………………………………………… 048

第二部分
TPM1～TPM19 联合公报

中日韩三国环境科学研究院（所）长会议第一次会议联合公报 ·················· 052

The First Tripartite Presidents Meeting among CRAES,

NIES and NIER Joint Communiqué ················· 054

中日韩三国环境科学研究院（所）长会议第二次会议联合公报 ·················· 056

The Second Tripartite Presidents Meeting among NIES,

NIER and CRAES Joint Communiqué ················· 058

中日韩三国环境科学研究院（所）长会议第三次会议联合公报 ················· 061

The Third Tripartite Presidents Meeting among NIER,

CRAES and NIES Joint Communiqué ················· 064

中日韩三国环境科学研究院（所）长会议第四次会议联合公报 ·················· 067

The Fourth Tripartite Presidents Meeting among CRAES,

NIES and NIER Joint Communiqué ················· 069

中日韩三国环境科学研究院（所）长会议第五次会议联合公报 ·················· 072

The Fifth Tripartite Presidents Meeting among NIES,

NIER and CRAES Joint Communiqué ················· 074

中日韩三国环境科学研究院（所）长会议第六次会议联合公报 ·················· 076

The Sixth Tripartite Presidents Meeting among NIER,

CRAES and NIES Joint Communiqué ················· 078

中日韩三国环境科学研究院（所）长会议第七次会议联合公报 ·················· 080

The Seventh Tripartite Presidents Meeting among CRAES,

NIESand NIER Joint Communiqué ················· 082

中日韩三国环境科学研究院（所）长会议第八次会议联合公报 ·················084

The Eighth Tripartite Presidents Meeting among NIES,
NIER and CRAES Joint Communiqué ·················086

中日韩三国环境科学研究院（所）长会议第九次会议联合公报 ·················088

The Ninth Tripartite Presidents Meeting among NIER,
CRAES and NIES Joint Communiqué ·················090

中日韩三国环境科学研究院（所）长会议第十次会议联合公报 ·················092

The Tenth Tripartite Presidents Meeting among CRAES,
NIES andNIER Joint Communiqué ·················094

中日韩三国环境科学研究院（所）长会议第十一次会议联合公报 ·················096

The Eleventh Tripartite Presidents Meeting among NIES,
NIER and CRAES Joint Communiqué ·················098

中日韩三国环境科学研究院（所）长会议第十二次会议联合公报 ·················100

The Twelfth Tripartite Presidents Meeting among NIER,
CRAES and NIES Joint Communiqué ·················102

中日韩三国环境科学研究院（所）长会议第十三次会议联合公报 ·················104

The Thirteenth Tripartite Presidents Meeting among CRAES,
NIES and NIER Joint Communiqué ·················109

中日韩三国环境科学研究院（所）长会议第十四次会议联合公报 ·················115

The Fourteenth Tripartite Presidents Meeting among NIES,
NIER and CRAES Joint Communiqué ·················119

中日韩三国环境科学研究院（所）长会议第十五次会议联合公报 ·················124

The Fifteenth Tripartite Presidents Meeting among NIER,
CRAES and NIES Joint Communiqué ·················127

中日韩三国环境科学研究院（所）长会议第十六次会议联合公报 ·················131

The Sixteenth Tripartite Presidents Meeting among CRAES, NIES and NIER Joint Communiqué ················· 135

中日韩三国环境科学研究院（所）长会议第十七次会议联合公报 ················· 140

The Seventeenth Tripartite Presidents Meeting among NIES, NIER, and CRAES Joint Communiqué ················· 144

中日韩三国环境科学研究院（所）长会议第十八次会议联合公报 ················· 149

The Eighteenth Tripartite Presidents Meeting among NIER, CRAES and NIES Joint Communiqué ················· 153

中日韩三国环境科学研究院（所）长会议第十九次会议联合公报 ················· 158

The Nineteenth Tripartite Presidents Meeting among CRAES, NIES, and NIER Joint Communiqué ················· 162

附录 本书缩略语中英文对照表 ················· 168

第一部分

TPM 机制发展回顾
与展望

1 TPM 机制建立背景、历次会议主要成果及会议举办方式

1.1 TPM 机制建立背景

为解决日益突出的本国及区域大气环境污染问题，中国环境科学研究院（以下简称中国环科院）、日本国立环境研究所和韩国国立环境科学院从 20 世纪 90 年代开始，开展"东北亚大气污染物长距离跨界输送项目（LTP）"合作，自此，建立了稳定的科研合作关系。1999 年，在东盟与中国、日本、韩国（10+3）领导人会议期间，中国、日本、韩国（以下简称中日韩）三国领导人首次举行三边会谈，提出了加强环境合作和对话的倡议。同年，中日韩三国举行了首次中日韩环境部长会议（TEMM），自此建立了东北亚环境领域的区域性高级别合作机制。在此背景下，中日韩三国政府部门、科研院所和民间社会开展了多层次、多领域的环境合作。

改革开放以来，我国环境保护意识日益加强。1978 年 12 月 31 日，中国环科院正式成立。1983 年 12 月，第二次全国环境保护会议召开，确立环境保护为我国必须长期坚持的一项基本国策。1996 年 7 月，第四次全国环境保护会议召开，

提出保护环境的实质就是保护生产力。1997 年 9 月，党的十五大报告强调，在现代化建设中必须实施可持续发展战略。2000 年，国务院印发了《全国生态环境保护纲要》，标志着我国生态环境保护工作进入了一个新的发展时期。中国环科院作为国家级的环境科研机构，立足于开展前瞻性、适用性的研究，为国家环境管理提供坚实可靠的科技支撑，在新的发展时期，需要学习和研究发达国家同类科研机构的运行管理机制，进一步提高科技创新能力。考虑到日本、韩国与中国地缘相近、文化相通，中国环科院以贯彻 TEMM 会议合作框架为契机，以学习借鉴日本国立环境研究所和韩国国立环境科学院经验为最初目的，历时 2 年的多方努力，于 2004 年 2 月联合日方与韩方，成功在北京召开第一次中日韩三国环境科学研究院（所）长会议（TPM1）。

TPM1 得到了当时的中国国家环境保护总局的大力支持，会上三国环境科研院（所）通过学术交流，进一步增信释疑，共同确立了中日韩三国环境科学研究院（所）长会议（TPM）机制，强调共同关注东北亚区域和次区域环境问题并开展联合研究。自此，在 TPM 框架下，中日韩三国环境科研院（所）每年轮流主办院长会议及其平行研讨会，其他两个科研院（所）长率团参会，至今已成功举办十九次会议。

尽管中日韩共处东北亚，但三国的国情差异较大，面临的环境问题与压力也各不相同，最终是三个科研院（所）开展环境研究的一致目标将三方聚集在了一起，并逐步确立了"友谊、交流、合作、共赢"的合作基本原则。中日韩三国环境科研院（所）不断在发展中完善优化 TPM 机制，围绕共同关注的环境问题开展科技交流与合作研究；针对本国和区域性重大环境问题，通过实验研究阐述环境污染问题的机理，通过交流开发控制污染排放、迁移和转化的技术，通过研讨为三国政府和环境主管部门制定有效的环境管理政策提供科学依据。20 年来，三国环境科研院（所）不断增进互信、深化合作、共同发展，**TPM** 也逐步成为中日韩三国环境领域合作的重要平台。

1.2　TPM 历次会议主要成果

表 1-1　TPM 历次会议主要成果

地点和日期	主要成果
TPM1 中国北京 （2004.02.16—2004.02.17）	▶ **机制建设：**建立 TPM 机制，就每年举行 TPM 达成一致； ▶ **交流内容：**回顾三方友好环境合作，交流各自管理、用人、研究、经费来源与财务管理机制等； ▶ **合作展望：**强调各国之间相互学习经验的重要性，讨论开展三方信息和人员交流以及加强合作的方式； ▶ **联合项目 / 合作：**探讨申请韩国国际协力机构 (KOICA) 援助 25 万 ~ 30 万美元建立长江三峡库区水环境研究实验室项目
TPM2 日本筑波 （2004.10.12—2004.10.14）	▶ **机制建设：**完善 TPM 机制，议定由年度院（所）长会议、不定期工作层会议和平行研讨会三部分组成； ▶ **交流内容：**确定 6 个优先研究领域，即淡水污染、大气污染（含移动源污染）、跨界大气污染、沙尘暴、有害物质污染、候鸟和湿地； ▶ **合作展望：**探讨开展人员交流，互派访问学者； ▶ **联合项目 / 合作：**议定由三方国际合作部门组织专家联合编制项目建议书：《湖泊富营养化调查规范》
TPM3 韩国济州岛 （2006.05.16—2006.05.17）	▶ **机制建设：**明确 TPM 机制，确定由年度院（所）长会议、平行研讨会和定期工作层筹备会组成；确立了优先研究领域联络人机制； ▶ **交流内容：**交流三方进行的院（所）科技创新机制改革；讨论三方在淡水环境、大气污染传输、沙尘暴等领域的合作进展和成果；

续表

地点和日期	主要成果
TPM3 韩国济州岛 （2006.05.16— 2006.05.17）	▶ **合作展望：** 就欢迎三方的兄弟研究院（所）的代表作为观察员参会达成共识，并考虑纳入东北亚地区其他国家的研究院（所）作为观察员参会； ▶ **联合项目/合作：** 回顾并肯定三方在"淡水污染防治项目"以及"东北亚跨界空气污染物"方面的合作
TPM4 中国成都 （2007.05.15— 2007.05.17）	▶ **机制建设：** 确立"友谊、交流、合作、共赢"的 TPM 合作基本原则； ▶ **交流内容：** 肯定 2006 年 12 月举办 TEMM 边会"区域生态系统及沙尘天气的影响和防治"国际研讨会成果； ▶ **合作展望：** 重申三方继续加强信息、出版物以及专家间的交流； ▶ **联合项目/合作：** 制定沙尘暴防治三方联合研究计划，同意在韩国为中国科研人员举办短期培训
TPM5 日本北海道 （2008.11.25— 2008.11.28）	▶ **机制建设：** 为"固体废物管理""气候变化"2 个新增优先研究领域指定联络人； ▶ **交流内容：** 交流优先研究领域合作进展及计划，讨论了三方人员交流计划； ▶ **合作展望：** 增加"固体废物管理""气候变化"2 个优先研究领域； ▶ **联合项目/合作：** 审议有关沙尘暴的联合研究项目，提出城市空气质量模型对比研究项目
TPM6 韩国首尔 （2009.11.25— 2009.11.27）	▶ **机制建设：** 同意 TPM 平行研讨会面向相关机构和专家开放； ▶ **交流内容：** 交流三方共同面临的全球环境挑战、优先研究领域活动进展； ▶ **合作展望：** 重申加强人员交流和信息共享；提议向中方派遣韩国顾问； ▶ **联合项目/合作：** 提出固体废物回收风险评估及安全准则的新合作项目建议

续表

地点和日期	主要成果
TPM7 中国青岛 （2010.09.12— 2010.09.16）	▶ **机制建设：** 认可在研讨会上与利益相关方分享专业知识能够带来切实的好处； ▶ **交流内容：** 就推进 8 个优先研究领域相关活动达成一致，将"候鸟和湿地"的优先研究领域重新命名为"生物多样性保护"； ▶ **合作展望：** 重申三方应继续进行信息、出版物以及专业知识的交流； ▶ **联合项目/合作：** 提议就跨界空气污染物、固体废物管理及其他问题开展项目合作研究
TPM8 日本冲绳岛 （2011.11.21— 2011.11.24）	▶ **机制建设：** 同意邀请对 TPM 优先研究领域感兴趣的其他国家相关研究人员和机构参加 TPM 会议； ▶ **交流内容：** 更新 8 个优先研究领域的名称、定义、内容； ▶ **合作展望：** 讨论灾害引发的环境问题； ▶ **联合项目/合作：** 强调 TPM 机制下加强具体项目合作的重要性
TPM9 韩国平昌 （2012.11.11— 2012.11.16）	▶ **机制建设：** 同意为每个优先研究领域指定牵头协调机构； ▶ **交流内容：** 突发环境危机应对； ▶ **合作展望：** 强调三方继续保持合作伙伴关系对应对东北亚地区面临的日益严峻的环境挑战的重要性； ▶ **联合项目/合作：** 无
TPM10 中国南京 （2013.11.04— 2013.11.08）	▶ **机制建设：** 调整 9 个优先研究领域的牵头协调机构；强调优先合作领域联络员应更积极参与 TPM 活动； ▶ **交流内容：** 三方院（所）研究与发展情况，增加"灾害性环境问题"（Disaster Environment）优先研究领域； ▶ **合作展望：** 灾害性环境问题影响研究； ▶ **联合项目/合作：** 无

续表

地点和日期	主要成果
TPM11 日本川崎 （2014.11.11— 2014.11.15）	▶ **机制建设：** 无； ▶ **交流内容：** 环境与健康、温室气体、环境应急、大气细颗粒物、生物多样性； ▶ **合作展望：** 三方务实合作对改善东北亚地区环境至关重要；赞同寻求与中日韩其他机构的科研人员开展合作；一致认为需进一步在优先研究领域开展合作； ▶ **联合项目 / 合作：** 分享"$PM_{2.5}$ 和短寿命气候污染物"专题研究进展
TPM12 韩国丽水 （2015.11.02— 2015.11.06）	▶ **机制建设：** 无； ▶ **交流内容：** 空气质量监测和预测系统、实际道路驾驶条件下的车辆排放控制、工业区公共卫生的环境流行病学调查、"未来地球计划"等新兴环境议题； ▶ **合作展望：** 强调优先研究领域的积极有效合作是延续三方成功合作的关键； ▶ **联合项目 / 合作：** 分享"气候变化"专题研究进展；审议了《TPM 优先研究领域合作路线图（2015—2019）》草案，并提出进一步修改完善建议
TPM13 中国昆明 （2016.10.31— 2016.11.04）	▶ **机制建设：** TPM 历史回顾、成果总结、现状评估、未来展望； ▶ **交流内容：** 环境基准与标准、中长期计划、气候变化、灾害性环境问题研究、沙尘和颗粒物； ▶ **合作展望：** 强调优先研究领域应基于各国环境研究优先次序，并将区域和全球环境问题纳入考虑； ▶ **联合项目 / 合作：** 通过了《TPM 优先研究领域合作路线图（2015—2019）》
TPM14 日本筑波 （2017.10.24— 2017.10.28）	▶ **机制建设：** 同意开展广泛、实质的 TPM 机制改革； ▶ **交流内容：** 环境研究数据库、环境健康、生态文明建设； ▶ **合作展望：** 同意进一步讨论优先研究领域合作弹性机制，开发三方合作研究项目； ▶ **联合项目 / 合作：** 决定 TPM15 审议《TPM 优先研究领域合作路线图（2015—2019）》实施报告

续表

地点和日期	主要成果
TPM15 韩国釜山 （2018.10.29— 2018.11.02）	▶ **机制建设：** 同意 TPM 改革方案，并于 TPM16 开始实施； ▶ **交流内容：** 环境健康；明确将 TPM 的 9 个优先研究领域调整为 4 个潜在研究领域，即大气污染、水环境、气候变化、环境健康； ▶ **合作展望：** 讨论培养青年研究人员和加强科学家合作交流的机制； ▶ **联合项目/合作：** 审议《TPM 优先研究领域合作路线图（2015—2019）实施报告》
TPM16 中国杭州 （2019.10.28— 2019.11.01）	▶ **机制建设：** 以 TPM 框架下的具体合作成果为重点推进 TPM 改革； ▶ **交流内容：** 东北亚区域大气污染物跨境传输研究、气候变化、环境污染和全球环境问题、细颗粒物、淡水中的藻华、微塑料和新兴污染物； ▶ **合作展望：** 青年研究人员和科学家需更加积极地参与到 TPM 中；同意共同发展"TPM+"模式； ▶ **联合项目/合作：** 讨论 4 个潜在研究领域的联合研究项目提案
TPM17 NIES 主办—线上会议 （2020.12.16）	▶ **机制建设：** 因新冠疫情，首次线上举办 TPM，未举办平行研讨会； ▶ **交流内容：** 气候变化、环境卫星、环境健康、大气污染控制、水环境修复； ▶ **合作展望：** 一致认可人工智能（AI）技术在环境科学中的重要性；讨论了 TPM 与其他合作机制或平台之间的关系； ▶ **联合项目/合作：** 中方倡议开展"中日韩三国环境科研院（所）科技创新体制机制对比研究"
TPM18 NIER 主办—线上会议 （2021.11.04）	▶ **机制建设：** 设置气候变化主题研讨，未举办平行研讨会； ▶ **交流内容：** 气候变化适应、人工智能（AI）和大数据、环境健康、环境卫星大气环境监测、废弃物管理、"双碳"目标等；

续表

地点和日期	主要成果
TPM18 NIER 主办—线上会议 （2021.11.04）	▶ **合作展望：** 认可三方院（所）的紧密合作对于实现低碳和可持续社会至关重要； ▶ **联合项目/合作：** 开展"中日韩三国环境科研院（所）科技创新体制机制深入对比研究"
TPM19 CRAES 主办—线上会议 （2022.11.24）	▶ **机制建设：** 以线上会议形式恢复平行研讨会的举办惯例；调整潜在研究领域牵头协调机构； ▶ **交流内容：** 气候变化、"双碳"目标、绿色低碳发展、环境健康、大气污染防治等； ▶ **合作展望：** 充分肯定和赞赏"对比研究"机制； ▶ **联合项目/合作：** 开展"中日韩三国环境科研院（所）气候变化领域对比研究"

1.3　会议举办方式

根据 TPM1（2004 年 2 月）的会议成果及后续对 TPM 机制的逐渐完善，TPM 会议由院（所）长会议、平行研讨会以及实地调研交流三部分构成。

（1）院（所）长会议

院（所）长会议，即 TPM 主会（TPM Main Meeting），内容主要包括三方科研进展报告、现有合作进展报告和未来合作讨论三部分。科研进展报告主要由三方分别介绍各自最新的院（所）发展情况、科技创新成果及发展规划等；现有合作进展报告一般包括三方在 TPM 优先研究领域（或潜在研究领域）的合作研究进展报告、TPM 平行研讨会会议成果报告，也可根据需要针对三方共同感兴趣的环境问题安排特别专题报告；未来合作讨论是由三国环境科研院（所）长围绕 TPM 未来发展、三方战略合作重点、共同关注问题等交换意见。会议成果在共同签署的联合公报中体现。

为进一步聚焦合作交流的环境领域，在 TPM2 上三方达成共识，提出了优先研究领域（Priority Research Areas），即在 TPM 框架下确定的共同感兴趣并可优先开展合作研究的领域。随后，根据三方共识，为促进三方进一步的实质合作，在 TPM15 上对优先研究领域进行了整合，将前期合作开展的 9 个优先合作领域调整为 4 个潜在研究领域（Potential Research Areas），并延续至今（表 1-2）。

表 1-2　TPM 优先及潜在研究领域发展情况

时间	优先及潜在研究领域（PRAs）
TPM2~TPM4 （2004—2007 年）	6 个优先研究领域： ①淡水污染（Freshwater Pollution） ②大气污染（含移动源污染）（Air Pollution including Vehicular Sources） ③跨界大气污染（Transboundary Air Pollution） ④沙尘暴（Yellow Sand Storm） ⑤有害物质污染（Hazardous Materials Contamination） ⑥候鸟和湿地（Migratory Birds and Wetland）
TPM5~TPM6 （2008—2009 年）	8 个优先研究领域（新增 2 项）： ⑦气候变化（Climate Change） ⑧固体废物管理（Solid Waste Management）
TPM7 （2010 年）	8 个优先研究领域： 将"候鸟和湿地"（Migratory Birds and Wetland）更名为"生物多样性保护"（Biodiversity Conservation）

续表

时间	优先及潜在研究领域（PRAs）
TPM8~TPM9 （2011—2012 年）	8 个优先研究领域的名称进行了更新： ①淡水污染（Freshwater Pollution） ②亚洲大气污染（Asian Air Pollution） ③城市环境和生态城市（Urban Environment and Eco-city） ④沙尘暴（Dust and Sand Storm） ⑤化学品风险及管理（Chemical Risk and Management） ⑥生物多样性保护（Biodiversity Conservation） ⑦固废管理（Solid Waste Management） ⑧气候变化（Climate Change）
TPM10 （2013 年）	9 个优先研究领域（新增 1 项）： ⑨灾害性环境问题（Disaster Environment）
TPM11~TPM14 （2014—2017 年）	9 个优先研究领域（保持不变）
TPM15 （2018 年至今）	TPM 进行改革，将 9 个优先研究领域改为 4 个潜在研究领域： ①大气污染（Air Pollution） ②水环境（Water Environment） ③环境健康（Environmental Health） ④气候变化（Climate Change）

　　为推动 PRAs 具体的合作交流，TPM 确立了 PRAs 联络员（FP）机制，每个领域三方各确定一位联络员。此外，在 TPM9 上，院（所）长们同意在每个优先合作领域选定一个牵头协调机构（LI）。牵头协调机构的主要职责包括：组织信息交流、培训，举办各类型研讨会，组织开展新合作项目等。在 TPM9 上首次确定了 LI 分工，在 TPM10 上确定增加"灾害性环境问题"为优先合作领域后，对 LI 的分工进行了调整，并一直保持到 TPM14。在 TPM15 上将原来的 9 个优先研

究领域调整为 4 个潜在研究领域后，调整确定了新的 LI。在 TPM19 上，院（所）长们商定每 3 年轮换一次各领域的 LI，并确定了第 1 次轮换方案从 TPM20 开始执行（表 1–3）。

表 1–3　PRAs 牵头协调机构 LI 分工情况

时间	CRAES	NIES	NIER
TPM9（2012 年）	淡水污染、化学品风险与管理、生物多样性保护	亚洲大气污染、城市环境与生态城市、气候变化	沙尘暴、固体废物管理
TPM10~TPM14（2013—2017 年）	淡水污染、城市环境与生态城市、化学品风险与管理	生物多样性保护、气候变化、灾害性环境问题	亚洲大气污染、沙尘暴、固体废物管理
TPM15~TPM19（2018—2022 年）	水环境	气候变化	大气污染、环境健康
TPM20（2023 年）	大气污染	气候变化、环境健康	水环境

（2）平行研讨会

TPM 平行研讨会上三方围绕共同感兴趣的环境问题，就研究进展进行分享交流，共同探讨解决方案，挖掘合作研究主题。其中，以大气环境和水环境为主要交流议题的分别占举办次数的 37.5% 和 25%。TPM 历年平行研讨会主题见表 1–4。

表 1-4　TPM 历年平行研讨会主题

时间	平行研讨会主题	涉及领域
TPM1	—	—
TPM2	淡水污染防治	水环境
TPM3	东北亚国家空气质量管理	大气环境
TPM4	大城市的空气污染，包括车辆废气	大气环境
TPM5	有害物质环境污染	化学品管理
TPM6	适应气候变化实现低碳社会	气候变化
TPM7	生物多样性保护与固体废物管理	生物多样性与固体废物管理
TPM8	亚洲大气污染和生物多样性保护	大气环境
TPM9	城市环境和生态城市，气候变化的影响与适应	气候变化
TPM10	淡水污染	水环境
TPM11	生态城市和生物多样性	生物多样性
TPM12	亚洲空气污染	大气环境
TPM13	水污染防治技术及生态系统健康	水环境
TPM14	通过评估和管理解决淡水环境问题	水环境
TPM15	固体废物管理与处置现状和前景	固体废物管理
TPM16	东北亚地区大气污染物长距离输送问题研究	大气环境
TPM17	—	—
TPM18	—	—
TPM19	基于新兴技术的大气观测与源解析	大气环境

注：受新冠疫情影响，TPM17 和 TPM18 未召开平行研讨会。

（3）实地调研交流

为促进中日韩三方更加深入地了解各方环境基础设施建设、环境治理方案，专设实地调研交流。TPM历次实地调研紧密围绕当时紧迫或关键的环境问题开展，具体情况见表1-5。

表1-5　TPM历次实地调研交流情况

时间	实地调研交流内容
TPM1	▶ 中国三峡库区环境设施
TPM2	▶ 日本国立环境研究所实验室（气候变化研究、内分泌干扰素实验室、土壤实验室） ▶ 日本足尾铜冶炼厂 ▶ 日本小田代原湿地
TPM3	▶ 韩国高山气象观测站 ▶ 韩国山地川生态修复
TPM4	▶ 中国四川省都江堰水利工程 ▶ 中国四川省大熊猫生态保护基地
TPM5	▶ 日本国立环境研究所大石岬全球环境监测站 ▶ 日本知床国立公园自然中心 ▶ 日本川汤生态博物馆 ▶ 日本摩周湖生态管理
TPM6	—
TPM7	▶ 中国青岛市生态工业园区
TPM8	▶ 日本冲绳边户岬大气环境监测站、冲绳山原野生动物保护中心、冲绳铁路中心 ▶ 日本热带生物圈研究中心、海洋博公园、琉球大学、冲绳卫生环境研究所

续表

时间	实地调研交流内容
TPM9	▶ 韩国江陵市（低碳绿色城市） ▶ 韩国汉江环境研究中心 ▶ 韩国国立环境科学院、仁川运河 ▶ 韩国首尔首都圈垃圾填埋厂
TPM10	▶ 中国江苏省蠡湖生态管理 ▶ 中国江苏省贡湖生态管理
TPM11	▶ 日本川崎市生态环境保护设施、生态工业园区、生命科学和环境研究中心 ▶ 日本富士吉田市生态环境保护设施、日本国立环境研究所环境观测站、日本自然生态保护公园
TPM12	▶ 韩国丽水国家工业园区 ▶ 韩国佳施加德士化工园区 ▶ 韩国顺天湾国家湿地生态园
TPM13	▶ 中国昆明市海东湿地 ▶ 中国昆明市石林国家地质公园
TPM14	▶ 日本国立环境研究所 ▶ 日本国立环境研究所保护及改善水环境观测与研究站
TPM15	▶ 韩国釜山广域市保健环境研究院 ▶ 韩国釜山绿色能源有限公司
TPM16	▶ 中国浙江省安吉县余村 ▶ 中国浙江省京杭大运河塘栖段水体生态修复
TPM17	—
TPM18	—
TPM19	—

注：受新冠疫情影响，TPM17、TPM18 和 TPM19 以线上会议形式举办，未开展实地调研交流。

2 TPM 机制发展历程

20 年来，在三国环境科研院（所）的共同努力下，TPM 机制经历了"从无到有，从有到优"的过程，日益发展成为三方开展科研交流、务实合作、互学互鉴、合作共赢的重要平台和有效机制，为区域环境质量改善提供了科技支撑、决策支持和智库服务。从聚焦本国自身环境问题，到凝聚共识助力提升区域环境质量和改善人民福祉，再到共同探索应对区域乃至全球环境问题，TPM 机制主要经历了合作起步、稳步拓展和务实深化三个发展阶段，呈现出不同特点。

2.1 合作起步阶段：TPM1~TPM4（2004—2007 年）

2.1.1 聚焦本国环境问题探索建立合作

TPM 机制建立初期，三方侧重解决本国自身环境问题。三国环境科研院（所）长们在 TPM1 时强调"各国之间相互学习经验的重要性"[①]，希望通过 TPM 机制

---

① TPM1 联合公报。

建立学习交流平台，提高各自应对本国环境问题的科技能力。这一阶段的发展特征体现在 TPM2 上确定的 6 个优先研究领域（PRAs）。其中，在淡水污染、大气污染（含移动源污染）和有害物质污染 3 个领域直观地体现了中日韩三国在经济高速发展时期所面临的主要环境问题。同时，三方也认识到解决国内环境问题离不开国际交流合作。以日方、韩方为主提出的跨界大气污染、沙尘暴、候鸟与湿地 3 个优先研究领域，反映了其国内在面临较为突出的大气污染、淡水污染和生物物种保护等环境压力下，开始关注周边国家对其生态环境影响。而中方在这一阶段尚以学习借鉴日本、韩国先进环境技术和有益经验为主。

2.1.2　以信息和经验交流为主要方式开启合作

在 TPM 机制起步阶段，三国环境科研院（所）的合作方式以信息和经验交流为主。TPM1 联合公报中提到"院（所）长们对如何开展三方信息和人员交流以加强现有合作进行了讨论"[①]；TPM2 联合公报中提到"三方探讨了未来在信息和人员交流及研究重点等领域的进一步合作活动"[②]。关于信息交流的具体方式，该阶段主要聚焦于各自环境科技发展的信息公开。例如，TPM3 联合公报指出"在信息交流方面，院（所）长们同意指定各自院所的国际合作负责人作为 TPM 机制联络人负责信息交流，英文版年度报告、重大研究项目清单和主要内容将存放于三国环境科研院（所）的图书馆以促进研究合作"[③]；在 TPM4 上"院（所）长们重申三国环境科研院（所）应当继续加强信息、出版物以及专家间的交流"[④]。

---

① 　TPM1 联合公报。
② 　TPM2 联合公报。
③ 　TPM3 联合公报。
④ 　TPM4 联合公报。

2.1.3 TPM 机制规则初步建立

TPM 机制规则为 TPM 后续发展奠定了重要基础。在 TPM1 时，"院（所）长们就建立三方定期沟通机制进行了讨论，就每年召开一次院（所）长会议达成共识"，并"一致认为应召开工作层会议，以讨论 TPM 会议和联合研究项目的具体问题"[①]；在 TPM2 上三方商定了"每年 TPM 会议期间要举办围绕特定议题的平行研讨会"[②]。在 TPM1～TPM4 这一阶段，三方确立了"友谊、交流、合作、共赢"的 TPM 合作基本原则，逐步建立年度 TPM 筹备工作层会、TPM 主会、平行研讨会及实地调研交流的机制规则。

2.1.4 "TPM+"理念初步提出和实践

"TPM+"的理念在 TPM 建立初期就受到三国环境科研院（所）长们的重视。例如，在 TPM2 上，院（所）长们讨论了"将中日韩三国的一些相关研究机构纳入 TPM 的可能性"[③]；在 TPM3 上"院（所）长们同意在未来的 TPM 会上邀请三国的同类研究机构作为观察员"，并提出"应寻求纳入东北亚地区其他国家（朝鲜、蒙古国和俄罗斯）的同类科研机构作为观察员参会"[④]。基于这些共识，中方率先在 2007 年中方主办的 TPM4 上，邀请了蒙古国国家气象水文环境监测局和中国中日友好环境保护中心的代表作为观察员参加了会议。

---

① TPM1 联合公报。
② TPM2 联合公报。
③ TPM2 联合公报。
④ TPM3 联合公报。

2.2　稳步拓展阶段：TPM5～TPM13（2008—2016 年）

2.2.1　兼顾本国和区域环境问题，面向全球环境挑战拓展合作

随着区域环境问题日渐凸显，三国环境科研院（所）也逐步转向对本国和区域性环境问题的"内外兼顾"。TPM5 在 6 个初始的 PRAs 基础上，新增"固体废物管理"为优先研究领域。同时，随着中日韩三国持续的经济增长，工业企业快速发展导致环境污染事件频发，环境风险管理需求日益突出。为加强对环境风险管理和控制方面的经验分享及科研交流，TPM10 新增"灾害性环境问题"（Disaster Environment）为优先研究领域。在这一时期，中国的环境管理得到极大发展。以《环境空气质量标准》（GB 3095—2012）发布为标志，中国环境管理开始由以控制环境污染为目标导向，向以改善环境质量为目标导向转变[①]，在生态环境的系统化协调管理上，与日方、韩方产生了更多互动。例如，在"固体废物管理"领域，三方开始共同探讨"3R"原则，即废弃物减量化（Reducing）、资源循环使用（Recycling）与废弃物再生利用（Reusing）下的循环经济理念；在"淡水污染"领域，三方开始交流水生态健康等系统化的环境治理理念和技术方法。

这一阶段，全球性环境问题日益受到国际社会的广泛关注，也逐步成为 TPM 机制下三国环境科研院（所）共同关心的议题。例如，在 TPM5 上韩方"表达了在'气候变化'领域深化合作的兴趣"[②]，促成"气候变化"成为新的优先研究领域，并在韩方主办的 TPM6、TPM9 上以"气候变化"为主题召开平行研讨会。

---

① 《国家环境保护标准"十三五"发展规划》。

② TPM5 联合公报。

日方在 TPM7 上提出"城市化、自然灾害和气候变化，是当今世界的紧迫任务"[①]，同时韩方提出"希望 TPM 框架下的活动不仅能为应对东北亚环境问题，也能为应对全世界范围的其他区域环境问题打下良好的基础"[②]。在 TPM13 上，中方进一步提出"在加强三国科研人员技术交流和沟通的同时，应放眼亚洲，关注与全球环境相关的'热点问题'"[③]。

2.2.2　合作方式由信息和经验交流向联合研究转变

这一阶段，三国环境科研院（所）的合作方式由早期的信息和经验交流为主转向开展联合研究以共同寻求环境问题的解决方案。例如，在"亚洲大气污染"领域，三方在这一时期开展了联合监测研究，并开始进行数据信息共享；在"大气污染（含移动源污染）"领域，在 TPM5 上院（所）长们"同意选择各国的一座城市作为城市空气质量模型研究的地点，并在沟通交流过程中形成研究计划"[④]；在 TPM6 上韩方提出"三方需汇聚力量和智慧，通过合作研究探寻应对环境挑战的具体解决方法，并在此过程中提高三方对区域环境问题的认识"[⑤]；在 TPM7 上中方提出"通过组织专题研究组，将三国环境科研院（所）的合作能力最大化"[⑥]。

PRAs 下的联合研究被视为技术和管理的核心要素。在 TPM 机制发展早期，每个 PRAs 的大部分活动都集中在三方共同关注的环境问题研究现状交流和成果

---

① TPM7 联合公报。
② TPM7 联合公报。
③ TPM13 联合公报。
④ TPM5 联合公报。
⑤ TPM6 联合公报。
⑥ TPM7 联合公报。

分享。随着交流合作的深入，三方充分意识到加强联合研究将为包括中日韩三国在内的东北亚地区提高环境质量带来积极成果。为推进联合研究，三方一致同意必须制订具体而有力的研究计划，为此在 TPM12～TPM13 三方共同制定了《TPM 优先研究领域合作路线图（2015—2019）》，明确了 9 个 PRAs 的研究愿景、目标和主要研究议题，通过共享环境解决方案，促进环境科技联合研究，共同应对新的挑战。

2.2.3 TPM 机制规则进一步完善

这一阶段，TPM 机制规则得到进一步完善，建立了优先研究领域的牵头协调机构（LI）机制（TPM9），为三国环境科研院（所）的合作拓展发挥了积极的促进作用。LI 的分工（表 1–3）主要依据三国环境科研院（所）的关注重点和研究优势而定。例如，"淡水污染"的 LI 一直由中方担任，这与中国在这一阶段持续大力开展水环境治理密切相关；而"亚洲大气污染""沙尘暴"等领域的 LI 由日方、韩方担任，反映出日方、韩方仍然保持着对区域性环境问题研究的主动性。

2.2.4 "TPM+"理念扩展

"TPM+"的理念始终贯穿于 TPM 的发展历程。除继续邀请三国环境科研院（所）以外的机构代表作为观察员参与 TPM 会议以外，这一阶段更以开放性的思路扩展了"TPM+"的理念。例如，在 TPM6 上"院（所）长们同意将研讨会面向中日韩三国的大学和研究所的环境领域专家公开"[①]；在 TPM8 上"院（所）

---

① TPM6 联合公报。

长们一致同意邀请 TPM 机制外对 TPM 八大优先合作领域感兴趣的其他国家相关研究人员和机构参与今后的 TPM 会议"[①]。在此共识基础上，TPM7、TPM8 的平行研讨会均实现了向会议举办地的有关机构和人员开放；TPM10 的平行研讨会向来自澳大利亚、荷兰等国家的机构和人员开放。

2.3 务实深化阶段：TPM14 至今（2017 年至今）

2.3.1 聚焦典型区域及全球性环境问题深化合作

随着中日韩环境科技水平与管理能力的持续提升，三国环境问题的解决均取得积极进展，经济社会发展步入新阶段，面临着新的发展要求，三国环境科研院（所）的科研重点均相应有所调整。2016 年，日本国立环境研究所发布了其第 4 个五年研究计划，在 TPM14 上日方在介绍其机构定位时指出，"日本国立环境研究所是开展日本本土和国际环境问题研究的核心机构，其研究活动涉及参与环境健康和气候变化相关大型长期项目"[②]。韩国国立环境科学院在 2015 年完成了其内部组织架构调整，形成了环境健康、气候变化与空气质量、水环境、环境资源、环境设施五大研究方向的研究部门组织架构。同期中国环科院也进入了全新发展阶段，基于中国生态文明建设总体战略要求，逐步成立了国家大气污染防治攻关联合中心、国家长江生态环境保护修复联合研究中心、国家黄河流域生态保护和高质量发展联合研究中心等。

---

① TPM8 联合公报。
② TPM14 联合公报。

在此背景下，三国环境科研院（所）从加强务实合作的角度出发，进一步聚焦新时期下三方共同关注的环境议题，将原先的 9 个优先合作领域调整为 4 个潜在研究领域，即大气污染、水环境、气候变化和环境健康。其中，"大气污染"主要考虑的因素除其区域传输的特性以外，还有其他区域关联性。大气治理对各国而言是共同受益的领域，即使各国大气治理方式和水平各有不同，但大气治理技术成果推广在区域内具有较为广阔的市场。另外，大气和气候问题密不可分，减污降碳协同是三国共同关注的环境科技发展领域。"水环境"的继续保留主要有三个方面的原因，一是水是生命之基，水环境是环境保护的重要抓手，是大气、土壤、生态多介质的主要传输纽带和媒介，具有重要研究意义；二是新时期水环境保护的概念和范畴拓展为水环境、水生态、水资源"三水"统筹，其过程机理和传输路径复杂，治理难度更大，很多科学问题还需要深入研究探索；三是中日韩三国地处东北亚地区，是一衣带水的邻邦，主要的连接媒介还是水（海洋），水环境保护是东北亚环境合作的区域热点。同时，TPM 放眼国际环境热点问题，在全球碳减排背景下，当前国际社会最为共同关注的"气候变化"，继续成为三方需要加强合作的焦点领域，体现出三方共同关注减碳合作的新未来。"环境健康"问题既是关系社会安定的重要民生问题，又是环境科学基础研究的前沿领域，面对全球环境与健康研究的快速发展，三方以健康风险防控为导向开展环境健康领域科研合作，有助于为健康风险控制政策的制定提供科学依据，为维护三国人民健康提供保障。

2.3.2 联合研究推动务实合作

"务实合作"成为 TPM 在新阶段深化合作的主要方向。例如，在 TPM14 上

"院（所）长们对优先合作领域下产出务实成果的必要性表示关切，并同意进一步讨论弹性机制，开发三方合作研究项目"[1]。在 TPM16 上，三国环境科研院（所）相关人员共同商讨，提交了 4 个潜在研究领域的项目提案，"三国环境科研院（所）长认为，推进 TPM 改革应以 TPM 框架下的具体合作成果为重点，循序渐进地进行"[2]。这一阶段，三方合作已不局限于信息交流和联合研究倡议，而且针对特定的研究领域开展了务实的合作研究。例如，在"大气污染"领域，三方共同开展了空气质量观测技术、模拟模型研究；在"水环境"领域，共同开展了流域水生态保护研究；在"气候变化"领域，聚焦温室气体排放观测开展了三国的对比研究；在"环境健康"领域，共同探讨了出生队列、暴露参数等的研究。这一时期的合作，中日韩三方之间互动积极、成效显著。其中，在中国大力开展"深入打好污染防治攻坚战"、实现"碳达峰、碳中和"国家目标的背景下，中方在大气污染、气候变化领域的合作交流更为积极主动。

2.3.3 TPM 机制改革优化

在三国环境科研院（所）均进入新发展阶段的背景下，TPM 也进行了机制发展的改革优化。三国环境科研院（所）长们在 TPM14 上就改革方向进行了商议，"同意开展广泛、实质的 TPM 机制改革，包括寻求 TPM 潜在的新角色，以便在加强现有研究合作的同时，进一步推动三方合作"[3]。在 TPM15 上确定的改革内容包括"缩短 TPM 会期；重组 TPM 日程及其平行研讨会；由联络员定期安排

---

[1]　TPM14 联合公报。
[2]　TPM16 联合公报。
[3]　TPM14 联合公报。

优先合作领域学术会议；为年轻科学家 / 研究员提供更多机会参加 TPM 活动并展示研究成果"①等，以提高目标导向的交流效率和结果导向的合作效力。

在这一时期，为进一步加强三国环境科研院（所）的交流互鉴，深化合作研究，在 TPM17 工作层会议上，中方倡议开展相关议题的"对比研究"。目前，"对比研究"已经从最初的科技创新体制机制对比研究拓展到具体合作研究领域的对比研究，在 TPM19 上院（所）长们一致认为"'对比研究'是探索三方共同关注的问题和寻求 TPM 机制下进一步务实合作的有效途径"②。

2.3.4 "TPM+" 理念继续深化

"TPM+"的理念在这一时期获得三国环境科研院（所）更为深刻的认同，主要是基于在共同面对诸如沙尘暴、气候变化等区域性和全球性环境问题时，三方均认识到需要各国携手努力、共同应对。在前期实践的基础上，三国环境科研院（所）长在 TPM15 上达成共识，"全力开展务实合作，共同开发'TPM+'模式"③。在 TPM16 上，三国环境科研院（所）长再次确认了"TPM+"的发展方向，"通过'TPM+'模式下科研项目和能力建设的统筹规划，促进 TPM 为改善东北亚区域环境质量贡献科技力量"④。

综上所述，经过近 20 年的发展，TPM 机制已成为东北亚区域环境合作的重要平台，为区域环境质量改善发挥了积极作用。其间，三国环境科研院（所）的科研与管理水平也日益进步，基于各自院所及国家环境事业需求的契合点，三方

---

① TPM15 联合公报。
② TPM19 联合公报。
③ TPM15 联合公报。
④ TPM16 联合公报。

的交流合作始终以互信互谅为前提开展，也达成了越来越多关乎区域乃至全球可持续发展的共识，充分发挥各自优势，为区域和全球环境治理与改善提供了科技支撑。同时，TPM 机制也是中日韩三国不断加强沟通、理解、互信的重要窗口，为三国维护团结稳定、改善区域环境质量、提升人民福祉、振兴亚洲经济繁荣贡献了重要力量。

3 科研领域合作进展

中日韩三国环境科研院（所）自 1995 年开展"东北亚大气污染物长距离跨界输送项目（LTP）"合作，至今已有近 30 年历史。

三国环境科研院（所）通过 TPM 机制不断加深相互理解，并基于三国经济社会发展阶段、主要面临的环境问题和三国环境科研院（所）重点科研工作，共同确定并及时调整科研合作的重点领域，开展了多样化的合作研究和交流活动，成果丰硕。

3.1 大气环境领域

3.1.1 合作背景

长期以来，大气环境污染问题一直是东北亚地区最为典型的区域环境问题之一，特别是大气污染物的跨界传输持续受到中日韩三国的高度关注。考虑到大气环境污染的区域传输性和严重性，结合中日韩三国环境科研院（所）在"东北亚大气污染物长距离跨界输送项目（LTP）"上的合作基础，"跨界大气污染"成为 TPM 成立初期最早确定的优先研究领域之一。

3.1.2　合作进展

①起步期。正如前述分析，在合作早期阶段，日方、韩方更多以解决其本国环境问题为出发点探究区域大气污染的起源与传输，中方在相关研究中尚处于"跟跑"状态。三方合作形式以信息交流为主。

②深化期。基于中日韩三方共同开展的大量科研活动及成果，合作形式逐步超越信息交流，更多侧重于共同探索区域环境问题的解决方案。这一阶段，随着中国先后出台《大气污染防治行动计划》《打赢蓝天保卫战三年行动计划》等文件，多措并举全力开展大气环境治理，将"改善空气质量"列入《中日韩环境合作联合行动计划（2015—2019）》的优先合作领域，TPM 机制也继续通过联合研究和设备共享等途径深化大气环境领域的合作。2015 年，中国环境科学研究院与韩国国立环境科学院签署合作协议成立"中韩大气联合研究工作组"，在空气质量观测、化学机理研究、模型模拟方面持续开展合作研究，每年召开研讨会交流研究进展，取得积极成效。中国环境科学研究院与日本国立环境研究所也在这一时期实现仪器设备等资源共享，在跨界大气污染传输观测等项目中进一步加强合作。

③创新期。近年来，随着东北亚区域环境空气质量得到显著改善，在 TPM机制下大气领域合作逐渐朝着更加深入、精细、引领的方向发展，开辟出大气污染物协同作用机理、空气质量模型模拟与预报、基于卫星遥感等新兴科技下的数据应用、减污降碳协同减排技术等新合作方向。在 TPM16 上三方系统梳理介绍各自在 $PM_{2.5}$、SO_2 及 VOCs 等大气污染物监测、排放清单、空气质量模型模拟以

及减排管控等方面的最新研究和举措，在此基础上，三方将"提升空气质量模型在大气模拟和预报中的准确度"确立为联合研究主题。在 TPM19 上，三方对研究成果进行了总结汇报，就基于地面观测和卫星数据，使用模式反演、数据同化和深度学习等技术改进大气模拟精度等进行了深入交流，有效提升了各自空气质量模型模拟的技术水平。同时，三方达成共识，下一步将聚焦减污降碳协同增效技术方向，开展重点行业 NO$_x$、VOCs 等污染物与温室气体协同减排关键技术的交流与合作研究。

3.2　水环境领域

3.2.1　合作背景

在 TPM 机制成立之初，中日韩三国均面临较为严峻的水环境污染问题。2004 年，中国 Ⅰ～Ⅲ类水质断面的比例为 41.8%，而劣Ⅴ类的水质断面占比达 27.9%[①]；韩国对 40 个湖泊进行调查显示，60% 的湖泊水质已超过其Ⅲ类水质标准，韩国政府从 2004 年开始实行水质污染总量管理制度[②]；日本虽采取了诸多措施，但其 2004 年湖泊和水库的化学需氧量（COD）指数符合环境质量标准的也仅占 50.9%[③]。在中日韩三国均面临水环境保护与治理压力的背景下，"淡水污染"成为三方共同关注的环境问题。TPM2 即以"淡水污染防治"为主题举办平行研

---

① 《2004 年中国环境状况公报》。
② 刘芳，王圣瑞，李贵宝，等 . 韩国湖泊水污染特征与水环境保护管理［J］. 水利发展研究，2015（6）：5.
③ 徐开钦，齐连惠，蛯江美孝，等 . 日本湖泊水质富营养化控制措施与政策［J］. 中国环境科学，2010（S1）：6.

讨会，"淡水污染"也被确定为 TPM 优先合作领域之一。

3.2.2 合作进展

①以"水专项"为契机开启水环境领域经验分享与交流。自 2006 年开始，中国环科院参与了中国国家水体污染控制与治理科技重大专项（以下简称"水专项"）的总体设计及部分重点项目课题研究。以此为契机，中日韩三国环境科研院（所）进一步加强在"水环境"领域的合作交流，分别在 2013 年、2015 年组织召开了专题研讨会，并于 2013 年、2016 年、2017 年举办了以水环境问题为主题的 TPM 平行研讨会，就水体富营养化控制、水生态系统修复、流域水环境管理、水质基准与风险评估等研究进展进行了交流。

②深化流域水环境管理交流分享。随着中国出台长江经济带发展战略、黄河流域生态保护和高质量发展战略，中国环科院协调开展了"长江生态环境保护修复联合研究""黄河流域生态保护和高质量发展联合研究"两大国家级联合研究项目，并推动 TPM 平台开展流域水环境管理的学术交流与对话。例如，TPM16 以流域水环境管理为主题，中方分享了中国长江生态环境保护与水系统模拟识别研究进展，韩方分享了其最大日负荷总量（TMDL）政策在韩国重点流域的实施，日方则主要介绍了其在流域污染物的长期监测与分析上取得的重要成果。

③聚焦水系统模拟预测、大数据以及生物技术等，深化交流先进科技在水环境治理领域的研究与应用。在 TPM19 上，韩方介绍了其在水系统模拟、水量水质高精度耦合模型、环境 DNA 快速检测和诊断技术等方面开展的研究应用，日方介绍了其开展诸多大尺度流域气候变化和水生态系统交叉学科研究进展，中方则介绍了应用环境 DNA 技术、水生态调查等手段，开展长江、黄河等重点流域

水生态系统演变研究的情况。

3.3　气候变化领域

3.3.1　合作背景

　　"气候变化"是当今国际社会共同关心的重大全球环境挑战之一，中日韩三国均高度重视并积极采取措施应对气候变化。2008 年 8 月，韩国将"低碳绿色增长战略"作为韩国新的远景目标。2009 年韩国政府发表了《绿色增长国家战略及五年计划》，将"应对气候变化及实现能源自立"作为主要战略目标之一，并相继公布了《国家能源基本计划》《低碳绿色增长基本法》等，实质性推进温室气体减排 ①。日本在应对气候变化方面也起步较早。2004 年，日本环境省设立了全球环境研究基金，集中高校、研究机构和企业的力量研究日本走向 2050 年低碳社会的发展路线图。2007 年，日本确立了"21 世纪环境立国"战略；2008 年，日本内阁会议通过了《实现低碳社会行动计划》，进一步把应对气候变化上升为日本的国家战略 ②。中国高度重视气候变化问题，于 1998 年成立了国家气候变化对策协调机构，并根据国家可持续发展战略要求，采取了一系列与应对气候变化相关的政策和措施，为减缓和适应气候变化作出了积极的贡献。2007 年中国发布《中国应对气候变化国家方案》，大力推进气候变化相关工作，并积极参与全球气候治理。在 2008 年举行的 TPM5 上，韩方表达了在"气候变化"领域

---

① 韩国的应对气候变化立法情况 [EB/OL]. https://www.cma.gov.cn/2011xwzx/2011xqhbh/2011xdt xx/201408/t20140825_258948.html, 2014–08–25/2023–07–11.
② 日本应对气候变化的"举国体制" [EB/OL]. https://www.wenmi.com/article/pyenn302nf61.html, 2022–09–26/2023–07–11.

深化合作的想法，得到中方、日方的积极响应，商定将其新增为 TPM 优先研究领域。

3.3.2 合作进展

①起步期。相较中方，日方、韩方在此阶段表现得更为积极主动。在韩国召开的 TPM6 上，举办了以"适应气候变化实现低碳社会"为主题的平行研讨会；之后在韩国召开的 TPM9 上，又举办了以"城市环境和生态城市，气候变化的影响与适应"为主题的平行研讨会，三方多次分享"气候变化"领域的研究成果。

②发展期。2015 年 12 月第 21 届联合国气候变化大会通过《巴黎协定》之后，"自下而上"的气候治理模式激发世界各国提出"国家自主贡献"减排目标，中日韩三方开展气候变化领域的合作也得到加强。在 TPM12 上三国环境科研院（所）长听取了"气候变化"领域的专题研究进展，"并一致认为气候变化是三国亟待共同解决的问题之一"[1]。随后，三国环境科研院（所）议定在 2015—2018 年"气候变化"领域的主要合作交流内容应包括：建立气候变化领域研究数据、政策、技术等共享机制；温室气体及短寿命气候污染物（SLCPs）排放分析及削减措施研究；应对气候变化的经济可行性政策研究等。在此基础上，TPM13～TPM15 对上述合作内容及研究成果进行了交流分享。

③深化期。随着中日韩三国分别提出各自的"碳达峰、碳中和"目标，TPM 机制下的气候变化合作得到进一步深化。2020 年 9 月 22 日，中国国家主席习近平在第 75 届联合国大会一般性辩论上宣布中国二氧化碳排放力争于 2030

---

① TPM12 联合公报。

年前达到峰值，努力争取 2060 年前实现碳中和。2021 年 5 月，日本国会参议院正式通过修订后的《全球变暖对策推进法》，以立法的形式明确了日本政府提出的到 2050 年实现"碳中和"的目标。2021 年 8 月，韩国国会通过《碳中和与绿色增长法》，使韩国成为第 14 个承诺到 2050 年实现"碳中和"的国家。在此背景下，TPM 机制下"气候变化"的交流合作越发积极，三方开始在温室气体观测与减排等方面共同探究新的技术方法。例如，TPM17 和 TPM18 上，中日韩三方就温室气体的测定分析方法、观测网络构建、数据信息平台建设、气候变化对环境空气质量的影响等进行了深入的交流分享；在 TPM19 上，由中方倡议，三国环境科研院（所）开展了温室气体观测、碳减排路径、气候变化影响及风险评估等领域的科研活动及成果合作对比研究。

3.4　环境健康领域

3.4.1　合作背景

日本在 20 世纪经历了富山骨痛病、熊本水俣病等环境公害事件后，通过一系列的环境立法加强环境保护，关注环境健康影响。日本重视并持续关注环境与健康的相关研究，在环境污染物低剂量、长时间暴露的毒性研究及危险度评价方法方面形成了其学术优势①。韩国也在经历经济快速发展和突出环境污染问题后，越发关注环境污染导致的直接或间接人体健康影响，于 2006 年颁布《环境健康 10 年综合计划》。2008 年韩国发布的《环境健康法》确立了以民众健康

———————————

① 于云江，全占军，汪丽娅. 日本环境健康研究概述. 环境污染与健康国际研讨会论文集. 2005：432–437.

为核心目标的立法理念，成为全球范围内第一部系统的环境与健康法 [①]。中国在2007 年发布环境与健康领域的第一份纲领性文件——《国家环境与健康行动计划（2007—2015 年）》，开启系统化的环境健康管理，2011 年发布《国家环境保护"十二五"环境与健康工作规划》，2016 年发布《"健康中国 2030"规划纲要》，开展重点区域、流域、行业的环境与健康调查。实施环境与健康风险管理成为中国建立健全环境与健康管理制度的重要举措。

在"环境健康"纳入 TPM 潜在研究领域之前，中日韩三国环境科研院（所）就已经在该领域分别开展了双边合作。例如，2012—2017 年间，在中韩环境合作联合委员会机制下，中国环境科学研究院与韩国国立环境科学院联合举办了六次"中韩环境健康论坛"，促进双方对环境健康相关法律法规和管理政策的相互了解，为进一步合作奠定了基础；2017 年，韩国国立环境科学院和日本国立环境研究所签署合作备忘录，针对儿童的环境健康开展联合研究。

3.4.2 合作进展

TPM15 将"儿童暴露于优先化学品的生物监测分析方法"确定为环境健康领域的联合研究主题。三方选择新型持久性有机污染物（POPs）全氟和多氟烷基物质（PFAS）为研究对象，对其生物监测分析方法进行了比较研究，并在 TPM16上进行了研究成果汇报。在 TPM17 时，三方开展了从胎儿到青少年 PFAS 暴露的研究交流，并对其他的 POPs 物质监测分析方法进行了比较研究。在 TPM19 上，三方进一步分享了在环境健康风险评估及指南导则编制、环境暴露参数及评价模

---

① 徐永俊，富贵，石莹，等.韩国《环境健康法》及对我国相关立法工作的启示 [J].环境与健康杂志，2016，33（2）：169–171.

型等方面的研究进展，同时计划在环境暴露参数确定方法领域开展进一步的合作研究。当前，相关工作正在稳步推进中。

3.5　其他科研领域

除上述领域以外，部分合作领域曾在 TPM 合作进程中发挥重要作用，但由于三国环境科研院（所）职能变化及 TPM 机制改革等因素，不再列为 TPM 潜在合作领域，转为以其他形式开展交流合作。

3.5.1　沙尘暴

自 20 世纪 90 年代开始，荒漠化进程在世界范围内加快，沙尘暴给东北亚地区带来了严重环境影响。基于三方对该问题的共同关注，在 TPM3 上，三国环境科研院（所）长就区域沙尘暴问题进行了深入讨论，提出考虑建立一个沙尘颗粒物三维监测系统，并一致同意全力支持三方合作开展沙尘暴联合研究，采取有效措施减少东北亚地区沙尘对环境的危害。为此，"沙尘暴"被确立为优先研究领域，也是 TPM 最早确定的优先研究领域之一。

随后，三国环境科研院（所）根据沙尘实际观测数据，利用化学传输模型（CTM）建立网格化数据集及亚洲沙尘预报系统，在提高沙尘预测的准确性、制定公共卫生和流行病学的保护措施方面进行了相关研究的交流。这些合作和交流为认识区域沙尘暴频发的主要原因、了解中国沙尘暴治理成效等起到了关键的增信释疑作用。后因沙尘暴发生频次强度有所缓解，沙尘暴问题被合并到大气环境领域。

自 2021 年 3 月以来，区域沙尘暴已进入新一轮频发期，利用和发展 TPM 机制，推进"TPM+"合作框架，联合蒙古国进一步加强沙尘治理技术研究与交流，是共同推进并解决这一区域重大环境问题迫在眉睫的行动需要。

3.5.2 生物多样性保护

"生物多样性保护"是中日韩三国共同关注的领域。在 TPM 成立之初，以"候鸟和湿地"为优先研究领域，相关合作以日方、韩方为主导，重点关注以白鹳为代表的珍稀候鸟在东北亚地区的栖息湿地生态系统的保护与研究。随着"生物多样性保护"在全球的不断深入推进，特别是 2010 年《生物多样性公约》第十次缔约方大会在日本名古屋召开后，三国环境科研院（所）在生物多样性保护领域的研究得到拓展，生物多样性监测与评估、外来物种入侵等主题受到三方共同关注，促使 TPM7 决定将"候鸟与湿地"更名为"生物多样性保护"。TPM8 和 TPM11 分别举行了以生物多样性保护为主题的平行研讨会；TPM12～TPM14，三方持续分享在生物多样性监测与评估、外来物种入侵等领域的研究进展。

随着三国环境科研院（所）研究职能的动态调整（生物多样性问题已不再属于日本国立环境研究所的研究范畴），生物多样性保护不再作为 TPM 机制的优先研究领域，但曾经在 TPM 机制下的交流对话为中日韩三国加强生物多样性保护合作奠定了良好基础。

3.5.3 固废管理

20 世纪 90 年代末，日本开始实施循环型社会建设，依托"3R"原则，成为世界上资源利用率最高的国家 ①。中国在《国家环境保护"十一五"规划》中进一步明确了以"减量化、资源化、无害化"为原则的"发展循环经济"的理念。通过研究安全处理和妥善管理废物以及扩大再生资源的利用来建立一个可持续的资源循环社会，是中日韩三国共同面临的环境问题和挑战，三国环境科研院（所）长基于此，在 TPM5 上，确定将"固体废物管理"纳入优先研究领域。

TPM7 以"固体废物管理"为主题召开了平行研讨会，TPM14 的主会上进一步设立了"废物管理"的特别议题，日方介绍了通过大量市政区垃圾收集的数据分析，对日本餐厨废物可堆肥比例进行的研究；韩方介绍了固体废物循环利用过程中开展环境风险评估的方法；中方介绍了中国危险废物污染现状、鉴别方法以及配套的标准体系。TPM15 再次举行"固体废物管理与处置现状和前景"主题平行研讨会，围绕"塑料废物管理政策与技术""废弃物能源化的管理现状和处理技术""固体垃圾衍生燃料与大气污染物排放管理"三个议题，三方分享各自研究进展。通过交流研讨，三方进一步认识到各自在该领域研究上的共同点和差异点，明确了未来合作交流的领域和方向。

3.5.4　化学品风险及管理、灾害性环境问题

日本、韩国在 20 世纪经历多次环境公害事件后，持续开展对有害物质的污染和环境风险研究。同时，随着中日韩三国持续的经济增长，工业企业快速发展，环境污染事件频发，环境风险管控需求日益突出。中国自"十一五"时期开始，逐步建立了"一案三制"的环境应急管理体系。在 2005 年松花江硝基苯污染事

---

① 　叶军 . 日本循环型社会建设路径研究：基于 3R 原则 [J]. 国外社会科学，2019（1）：40-52.

件后，中国突发环境事件应急管理进一步受到高度重视，从应急管控到技术应对均亟待加强完善。建立一个三边体系，通过基于科学和技术的解决方案有效应对潜在的灾害性环境问题成为三国环境科研院（所）长的高度共识。TPM2 确定"有害物质污染"为优先研究领域。在 TPM8 上三国环境科研院（所）长达成一致，将"有害物质污染"更名为"化学品风险及管理"。TPM10 将"灾害性环境问题"增加为优先研究领域。

"有害物质污染"为优先研究领域时，主要关注持久性有机污染物（POPs）及内分泌干扰物质等新型环境污染物。TPM5 首次举行了以"有害物质环境污染"为主题的平行研讨会。在"化学品风险及管理"成为优先研究领域后，三国环境科研院（所）围绕促进联合研究、对现有和新的化学品进行有效的评估和管理、建立化学品全生命周期的评估系统等议题进行了交流。"灾害性环境问题"成为优先研究领域后，三国环境科研院（所）在洪水等自然灾害、化学品泄漏等人为灾害引发的环境风险及污染应急管理等方面开展了一系列科研交流和经验分享。三方还曾计划围绕重大灾害性环境问题和确定重点灾害、建立灾害性环境问题及时信息共享系统、制定应急监测、预警系统和污染废物管理标准化措施等进行研讨。后因在 TPM15 上，TPM 进行优化改革，将 9 个优先研究领域改为 4 个潜在研究领域，"化学品风险及管理"和"灾害性环境问题"两个领域的交流未再延续进行。

3.6 小结

科研领域的交流合作是 TPM 发展的核心，在 TPM 机制下三方科研领域的合作意愿与共识不断增强，合作机制建设稳步完善，合作领域逐步拓展延伸，合作深度日益强化，合作水平不断提高，合作成效也日益提升。概括而言，TPM 科研合作主要呈现以下四方面的发展特点：

（1）互学互鉴，合作共赢共同提升

TPM 机制下的科研合作在尊重中日韩三国在国情、发展阶段、自然环境等方面各有差异的现实情况基础上，在求同存异中汇聚共识，在互学互鉴中互促提升。在 TPM 机制成立之初，日本、韩国整体的社会经济发展水平高于中国，其环境管理体系及环境科技研究水平整体超前于中国。在此背景下，TPM 机制下的科研合作，日方、韩方更多发挥带动、引领作用，中方多为学习、借鉴。随着中国社会经济的不断发展，特别是在习近平生态文明思想指引下，中国的生态环境管理逐步深入，生态环境质量改善成效显著。中方在环境科技研究上有了长足进步和发展，在 TPM 机制下由跟跑参与转变为主动作为、积极引领。同时，三方合作也在互学有益经验、互鉴最佳实践、互促能力提升的过程中形成资源共享、优势互补、合作共赢、共同提升的良好局面。

（2）携手共治，凝聚全球环境治理共识

TPM 机制从关注各国本身的环境问题，到积极交流区域环境问题治理，再到共同探索应对全球环境问题，三国环境科研院（所）不断凝聚共识，合作领域不断拓展延伸，合作方式不断引入新的科技方法。三方的早期合作主要以交流信息、分享经验及解决本国环境问题为出发点，如"大气环境""生物多样性保护"领域，日方、韩方更多关注大气污染区域输送等跨界环境问题对其国内生态环境质量的影响。随着气候变暖、生物多样性丧失等全球性环境问题日益突出，绿色低碳发展成为全球可持续发展的大势所趋。中日韩三方也逐步意识到凝聚共识、强化协作、共筑生态环境保护合力的重要性。因此，三国环境科研院（所）的合作从各

国国内问题逐渐聚焦到区域及全球性环境问题，合力推动全球生态环境治理技术和解决方案的研究及科研交流，TPM 机制为推进落实 2030 年全球可持续发展目标注入了东北亚区域合作的强大科技动力。

（3）与时俱进，技术赋能合作深化拓展

TPM 机制合作与时俱进，各领域合作从对传统科技手段的交流到对新兴科技综合应用的研究转变，合作范围和路径不断拓展。TPM 机制下早期的学术交流内容主要是采用传统的环境调查、监测分析等手段开展研究，随着科技的发展，综合应用新兴技术开展环境保护和综合治理的研究趋势愈加明显。例如，在"水环境"领域，早期的交流研讨侧重于各方对水质理化指标的监测和评估，而在近年则关注水生态系统调查与修复技术、环境 DNA 快速检测和诊断技术等在"水环境"管理中的应用；在"大气环境"领域，早期的交流重点在于采用传统监测手段开展大气污染物的监控，而近年来运用在线监测、无人机、遥感卫星等新兴技术手段获取大数据的信息技术在大气环境中的跟踪研究，成为三方合作研究的热点领域。

（4）前景广阔，环境科技合作大有可为

TPM 机制下的合作不仅具有良好的合作基础，也拥有广阔的合作前景。现阶段，气候变化、荒漠化、环境污染、生物多样性丧失等问题相互交织，全球环境面临的挑战越发严峻。在此背景下，TPM 机制下的环境科技合作意义重大，不仅对中日韩三国具有重要意义，也更具有世界意义。下一步，三方可在积极应对气候变化、治理环境污染、改善生态环境等领域继续加强交流，更加积极主动地开展更紧密、更务实的合作，为携手推进全球环境治理，推动构建人类地球命运共同体，共建清洁美丽的世界提供更为有效的科技支撑。

4 科研体制机制交流成果

在 TPM 机制发展的近 20 年中，三国环境科研院（所）的合作逐步由信息共享、科研交流延伸到政策研讨、机制交流等方面。其中，三国环境科研院（所）科研进展报告是 TPM 会议的重要内容之一，每年三方均分享各自最新的科技创新发展情况。在 TPM17 工作层会议上，中国环科院倡议开展系统的中日韩三国环境科研院（所）科技创新体制机制对比研究（以下简称对比研究），得到日方、韩方一致支持。随后，三方的对比研究范围从最初的科技创新体制机制对比，扩展至三方机构发展战略、科研管理和科技产出、资源配置以及分学科领域的深入对比，有效促进了三国环境科研院（所）科技管理方面的互学互鉴，而中方从科技体制机制的对比交流中也获益良多。

（1）明确战略发展定位，优化环境科技发展顶层设计

中国环境科学研究院和韩国国立环境科学院分别隶属中国和韩国的环境部，为环境管理、政策制定提供科技支撑是两个机构战略发展定位的共同点，而日本国立环境研究所作为独立运营机构（Incorporated Administrative Agency），更侧重于开展相对独立的环境科技基础研究。三国环境科研院（所）虽然在体制上有所差异，但均从各国生态环境领域的国家战略角度出发，聚焦国家重大环境战略需求，开展院（所）发展的顶层设计。

日本国立环境研究所从 2001 年开始制订了第 1 个五年计划，并分别在 2006 年、2011 年、2016 年和 2021 年制订了第 2 ～5 个五年计划。在 2021 年最新的五年计划中，日本国立环境研究所以环境问题和使命为导向，着力在低碳、资源可再生利用、环境健康、生物多样性等领域加强学科发展，以支持日本国家环境保护战略规划的实施。韩国国立环境科学院在发展过程中经历多次组织架构的调整，根据韩国国家重大环境战略需求，其当前发展愿景是"通过我们的研究与知识，为民众创造安全的环境及不断改善的环境质量"，其研究导向侧重于"碳中和和绿色发展政策"，其环境质量改善领域的研究目标是"民众满意度"，环境安全领域的研究目标是"适宜的生活环境"，基于此，气候变化与碳减排、资源循环利用、环境健康等领域是其学科发展的重点。

中国环科院在深入分析中国"十四五"时期（2021—2025 年）生态环境保护面临的新形势和新挑战基础上，结合自身发展优势和定位，制订了"十四五"科技发展规划，着力在碳减排、环境空气质量改善、重点流域生态保护、生态系统保护及修复等领域加强学科和科研平台建设，旨在为中国生态文明建设和绿色发展提供前瞻性的科技支撑。此外，中国环科院逐步推进"使命导向"科研院所管理改革，这与日本、韩国环科院（所）当前的发展理念同向而行，三方的交流互动更易积极有效。

（2）提升科研硬件能力，加强生态环境科研实验室建设

实验室是组织高水平科学研究、培养和集聚创新人才、开展学术合作交流、促进协同创新的重要载体和平台，TPM 机制下的中日韩三国环境科研院（所）也高度重视加强实验室能力建设方面的交流互鉴。

日本国立环境研究所的实验装备和设施达到日本国内一流水平，拥有光化学反应箱、自然环境模拟系统、淡水环境模拟系统和废物处理模拟系统等大批一流的科研设备，以及多个野外观测研究台站。日本国立环境研究所的实验室管理机制完善，科研作风务实严谨，利用强大的实验室能力，开展了大量跨学科、跨部门和跨领域的综合性研究，在日本环境保护和区域环境问题解决方面发挥了重要的作用。韩国国立环境科学院拥有 2000 余台主要的实验仪器设备，包括针对环境健康研究的动物实验设备、大型环境监测分析仪器、自动监测分析设备等；此外，韩国国立环境科学院下属的 10 个环境空气质量监测站、4 个流域研究中心及水质监测站，持续开展环境质量监测与数据信息积累工作。

中国环境科学研究院积极谋划加强生态环境科研实验室能力建设，现有 2 个国家级实验室（环境基准与风险评估国家重点实验室、湖泊水污染治理与生态修复技术国家工程实验室），9 个部级重点实验室及 4 个部级科学观测研究站。在国家和生态环境部科技体制机制创新与改革部署下，中国环境科学研究院按照国际化、标准化、系统化、智慧化建立了检测与实验大楼，总面积 7000m^2，实验室达 100 间，各类精密仪器设置上千台（套）。另外，依托中国环境科学研究院，中国生态环境部和韩国环境部共建了中韩联合环境研究实验室。在此过程中，同步推动了资源整合、硬件升级、人才汇聚及能力提升，大大提高了科研攻关的力度。

（3）提升科技创新软实力，强化科研创新平台建设及科研人才发展

科研创新平台建设和科研人才发展是科技创新软实力的重要体现。在 TPM机制下，中日韩三国环境科研院（所）在人才发展和平台建设方面持续深化交流。

在人才发展方面，三方达成共识，加强青年人员交流和人才培养力度，并通过在大气、水等领域的合作研究实现了科研人员的互访交流，极大地提升了三方人员科研能力。

在平台建设方面，日本国立环境研究所作为开展日本本土和国际环境问题研究的核心机构，其研究活动涉及环境健康和气候变化相关大型长期项目，并以这些项目为依托成立了日本环境与儿童研究以及气候变化适应中心，统筹相关领域的数据与研究。韩国国立环境科学院则整合资源着力在环境空气、气候变化、环境健康以及固废管理等领域开展了平台建设（如韩国国家应对气候危机中心等）。中国聚焦大气环境治理重点关键问题，由生态环境部牵头，以中国环境科学研究院为主要依托单位，组建了国家大气污染防治攻关联合中心，组织 269 家国家生态环境科研优势单位，2900 多名科技人员，联合开展研究（称之为"1+X"模式）。采用同样的模式，还组建成立了国家长江生态环境保护修复联合研究中心、国家黄河流域生态保护和高质量发展联合研究中心，用科技支撑中国生态环境质量持续改善。

5 面临的挑战与未来合作展望

在过去的 20 年里，中日韩三国环境领域的合作保持稳步有序发展，TPM 作为三国加强环境科技交流的平台，已成为一个区域合作的成功典范。当今世界正处于百年未有之大变局，随着国际形势深刻演变、世界经济新旧动能转换、科技革命与产业变革迅速发展，各国利益和命运也将前所未有地紧密相连、深度交融，中日韩三国更应加强合作，为东北亚地区和国际社会面对的广泛问题作出积极和应有的贡献。同时，全球环境治理已进入新的历史发展时期，TPM 机制下环境科技合作的务实深化和创新引领也将面临一系列新的挑战。针对这些挑战，对未来合作展望如下：

（1）机制层面：需要更好地引领区域合作，进而有效贡献全球环境治理

TPM 长期以来通过交流和联合研究解决了三国面临的许多关键环境问题。同时，三国环境科研院（所）也一直积极探索并努力寻找有助于改善区域环境质量的科学解决方案。三个研究机构一致认为，在优先研究领域开展积极有效的合作至关重要，并制定了 TPM 路线图，以促进相关环境研究。但系统性及规划性合作框架的缺失导致既有合作机制有如下问题：一是合作机制结构松散，缺乏及时沟通与协调，未能在应对气候危机和保障全球可持续发展的过程中形成合力；二是虽然在三边层次上致力于加强气候、大气环境、水环境等领域的科学合作，但尚未产生有效的政策联系；三是既有合作方式主要体现为召开会议，达成的共识性合作条款大多分散在声明性文件中，并不具备法律效力，难以推进区域环境合作实现实质性的共同决策和联合行动。因此，TPM 应立足于打造环境命运共同体的格局和理念、引领区域并推进全球环境治理技术和解决方案的研究和交流，提升合作战略的定位和共识。

（2）合作层面：需要进一步开展科技前瞻性合作，更好地应对复杂环境问题

三国环境科研院（所）从最初在环境科技领域的交流对话到逐步开展合作研究，合作成果丰硕。随着区域及全球环境治理进入"深水区"，亟待进一步深化环境科技合作。既有合作机制有如下问题：一是在合作广度上，继续集中在传统学科大气环境、水环境等领域开展软科学合作的问题突出，缺乏压倒性优势技术的研发合作，缺少开发设备化、智能化和信息化环境治理拳头产品的合作。二是在合作深度上，TPM 作为环境领域的科研风向标，缺乏一定程度的顶层设计和前瞻性战略规划，不能为共同应对更复杂的环境问题提供科技前瞻储备和前沿支撑。因此，TPM 应立足于国际科技前沿，持续加强人工智能、大数据融合、交叉学科等领域的前瞻性合作。

（3）体系层面：需要加强污染应急监测方面的协作，凝聚大安全大应急合力

中日韩三国一衣带水，自然生态环境相互依存，建立大安全大应急框架，对完善现代环境治理体系，严防环境和安全风险，保障东北亚乃至全球人民生命财产安全具有重要意义。在既有合作机制下，TPM 可以加强沟通交流频率，并建立联合应急监测和对突发环境事件的会商研判和信息共享。从科研合作层面促进中日韩三国建立真诚友善相互认知的良好氛围。同时，应加大 TPM 宣传力度，积极拓宽宣传渠道，使科学理论话题与人民日常生活对接，把握舆论引导的话语优势。

（4）资源层面：需要进一步开拓资金渠道，建立多要素科技创新生态

资金、技术、人才、数据等多要素的协同支撑是 TPM 开展进一步工作的实际需求。在既有合作机制下，影响 TPM 进一步开展实质性合作的一个重要障碍是缺少稳定的资金支持。三方主要借助各自国内相关科研项目资金，开展相对短期或小范围的联合研究或成果交流，针对性、系统性不足，尚未树立品牌性的联合研究项目，直接影响了 TPM 机制下更高质量、更具影响力的合作成果产出。因此，TPM 应基于三方已有研究基础和研究方向，共同谋划争取各类国际、国内资金资源的支持，发展促进科技项目合作的内在动力。

（5）地缘层面：需要进一步创新"TPM+"发展模式，持续保持发展动力

TPM 合作愿景是"改善东北亚人民的生活环境质量"。TPM 合作是东北亚区域环境合作最成熟的机制之一，长期以来为东北亚环境问题的解决和人民生活福祉的改善作出了积极贡献。在既有合作机制下，三国环境科研院（所）曾制订了具体合作研究计划，以确保每一项研究活动的成果可造福于区域更多国家环境问题的解决。中日韩三国作为区域主要经济体，与区域内其他国家面临诸多相似问题和挑战，TPM 理应带动拓展更多国家和科研机构参与合作，共迎挑战。"TPM+"是加强区域共治的必然发展趋势。尽管三方从初期就对"TPM+"模式进行了探索实践，但至今仍未在机制上产生实质性突破。吸引周边国家和相关机构参与 TPM 机制下的交流、合作，是保持 TPM 发展动力并引领区域及推进全球环境治理不可或缺的有效途径。

6 结语

　　TPM 机制从无到有、从有到优，成长为中日韩三国环境科研院（所）进行环境对话、科技交流、联合研究的重要平台和载体，也是中日韩三国环境科技领域交流合作最为成功、历史最久的合作机制。20 年来，TPM 合作机制逐步完善，达成的合作共识不断增加，内涵愈加丰富，形式持续创新，三国环境科研院（所）在共同关注的大气环境、水环境、气候变化、环境健康、沙尘暴等领域合作富有成效，为推动区域环境质量逐步改善、助力东北亚区域绿色发展作出了积极贡献，同时也为全球环境治理提供了宝贵经验。其间，无论国际形势如何变幻，三国环境科研院（所）始终秉持互信互谅、合作共赢的理念，保持了良好互动和友好合作，打造了互利共赢和引领未来的合作模式，成为促进区域环境质量持续改善的重要力量，切实发挥了科技服务外交的积极作用。

　　全球环境治理已进入一个新的历史时期，深化中日韩环境科技合作，对携手应对严峻的全球环境挑战，推动全球环境治理意义重大。真切希望 TPM 在新的历史时期，继续延承"友谊、交流、合作、共赢"的基本原则，立足引领区域并推进全球环境治理。相信在三方的共同努力下，TPM 将继续以科技推动绿色创新，以创新促进发展共赢，为东北亚区域生态环境绿色高质量发展发挥更大作用，为全球可持续发展作出更大贡献！

第二部分

TPM1~TPM19
联合公报

中日韩三国环境科学研究院（所）长会议第一次会议
联合公报

2004 年 2 月 17 日　中国北京

　　院（所）长们特别指出了三方过去几年间在环境科学技术领域的友好合作关系。院（所）长们还回顾了各自院所在环境科学技术方面的发展现状并分享了未来的研究计划。他们还从院（所）层面就如何提升环境科学技术交换了意见，强调各国之间相互学习经验的重要性。

　　院（所）长们呼吁三方要保持长期交流并拓宽合作领域，共同致力于改善国家、次区域和区域的环境。他们一致认为开展联合研究项目将有效促进三方对话与合作。他们认识到合作研究项目的规模主要取决于能够获得的资金规模，可通过向国际组织申请资助来扩大资金资源途径。

　　院（所）长们对于如何开展三方信息和人员交流以加强现有合作进行了讨论，并决定与包括国际科研组织在内的其他科研组织分享联合研究项目成果。

　　院（所）长们就建立三方定期沟通机制进行了讨论，并就每年召开一次院（所）长会议达成共识。下一届会议将由日本国立环境研究所主办。院（所）长们还讨论了未来在三国环境科研院（所）长会议期间，举办围绕特定议题的平行研讨会的可能性。他们一致认为还应召开工作层会议，以讨论关于三国环境科研院（所）长会议和联合研究项目的实际问题。

院（所）长们重申承诺，将继续保持联合磋商，以巩固加强三国环境科研院（所）长会议及相关活动所依据的合作框架。

The First Tripartite Presidents Meeting among CRAES, NIES and NIER Joint Communiqué

February 17, 2004, Beijing, China

The presidents noted the friendly relationship among CRAES, NIES and NIER in the area of environmental science and technology over the last few years. The presidents reviewed the present status of environmental science and technology in each institute and shared their future research plans. They also exchanged views on the institutional aspects in promoting environmental science and technology, stressing the importance of drawing lessons from experiences of each country.

The presidents called for sustained communication and expanded cooperation among the three institutes in pursuing their common research interests to improve the national, sub-regional and regional environment. They agreed that working on joint research projects would prove effective in promoting dialogue and cooperation. The presidents also recognized that the scope of cooperative projects largely depends on the availability of financial resources, and such resources might be augmented by seeking grants from international organizations.

The presidents discussed the exchange of information and personnel among three institutes to strengthen the existing cooperation. The presidents further decided to share the results of joint research projects with other scientific communities, including international ones.

The presidents discussed instituting a regular communication mechanism among the three sides. Their consensus was to hold the presidents meeting once a year. The next meeting will be hosted by NIES. They also discussed the possibility of holding a workshop on specific topics on the occasion of Tripartite Presidents Meeting in the future. They further agreed that a working level meeting should be convened to discuss practical matters of the presidents' meetings and joint research projects among the three institutes.

The presidents renewed their commitment to maintain joint consultations to reinforce the cooperative framework on which Tripartite Presidents Meeting is founded and its activities are conducted.

中日韩三国环境科学研究院（所）长会议第二次会议联合公报

2004 年 10 月 14 日　日本筑波

10 月 12 日举办了以"淡水污染防治"为主题的联合研讨会。该主题在中日韩三国环境部长会议（TEMM）上确立，被认定为东亚地区的重要环境问题，也是 TEMM 研究项目的主题之一。

在 TPM2 会议开幕式上，三国环境科研院（所）长特别强调了三方在环境科学技术领域的友好合作关系，并表示很高兴在 2004 年 2 月于北京召开的第一次中日韩三国环境科学研究院（所）长会议后再次会面。

院（所）长们指出 8 月在日本举行的工作层会议卓有成效，促进了 TPM2 的成功举办。他们一致认为未来应当延续召开工作层会议的惯例，并纳入对近期联合研究项目的相关讨论。

会上，三方介绍了各自目前的研究活动并探讨了未来在信息和人员交流以及研究重点等领域的进一步合作活动。

在信息交流方面，院（所）长们就在各自院所官方网站上增设 TPM 页面及定期交流出版物达成一致，具体形式将由三方的国际合作办公室通过邮件沟通决定。

在研究重点方面，院（所）长们一致同意在以下领域探索联合研究项目的可能性：

- ●淡水污染
- ●大气污染，包括移动源污染
- ●跨界大气污染
- ●黄沙尘暴
- ●有害物质污染，如内分泌干扰物及持久性有机污染物（POPs）
- ●候鸟与湿地

项目进展过程将由三国环境科研院（所）间的工作层会议制定。

院（所）长们进一步讨论了将中日韩三国的一些相关研究机构纳入 TPM，以扩大会议范围，并同意在选定未来工作层会议参会者时将这一点纳入考虑。

他们还讨论了加强三方合作的措施和行动，同意继续将每年一次的 TPM 会议机制作为定期交流的平台。韩国国立环境科学院院长提议于 2005 年秋在韩国举办 TPM3 及其平行研讨会。平行研讨会的主题将于 2005 年春举行的工作层会议确定。

院（所）长们对 TPM 的第一次平行研讨会表示满意，并赞同未来每年在召开 TPM 的同时，均围绕具体环境议题举办平行研讨会。

院（所）长们同意就各自的研究评估经验进行交流分享。

The Second Tripartite Presidents Meeting among NIES, NIER and CRAES Joint Communiqué

October 14, 2004, Tsukuba, Japan

The meeting was preceded by a joint workshop on "Freshwater Pollution Prevention Control" held on October 12th, a topic identified by the Tripartite Environment Ministers Meeting（TEMM）as an important environmental issue in East Asia and the subject of a TEMM research project.

At the TPM opening, the three presidents noted the friendly relations among NIES, NIER and CRAES in the area of environmental science and technology and expressed their pleasure to meet again following the First TPM held in February 2004 in Beijing.

The presidents further observed that the working level meeting held in August in Japan was highly fruitful to the success of the second TPM and agreed that the practice should be continued and expanded to include discussions of joint research projects in the near future.

During the meeting, the institutes described their recent research activities and discussed further cooperation activities on the topics of information exchange, personnel exchange and research priorities.

On the topic of information exchange, the presidents agreed to develop the TPM homepage in each institute web site and to exchange publications regularly, the precise modalities to be decided over e-mail by offices of international coordination of the three institutes.

On the topic of research priorities, the presidents agreed to explore possibilities for joint projects in the areas of:

● Freshwater pollution,
● Air pollution including vehicular sources,
● Transboundary air pollution,
● Yellow sand storm,
● Hazardous materials contamination, such as endocrine disrupting chemicals and POPs, and
● Migratory birds and wetland

Processes for project development will be formulated through working level meetings among the institutes.

The presidents further discussed expanding the TPM to include kindred research bodies in the three countries, and agreed to consider this in selecting future participants for working level meetings.

They also discussed further steps to strengthen the existing and evolving cooperation among NIES, NIER and CRAES, and agreed to continue the annual TPM as a vehicle for regular communication. President of NIER offered to host the Third TPM in Korea during autumn 2005, including a workshop whose theme will be decided at the working level meeting next spring.

The presidents expressed satisfaction with the first TPM workshop and agreed to convene future workshops on special environmental topics in conjunction with the annual TPM.

The presidents agreed to exchange their experiences of research evaluation.

中日韩三国环境科学研究院（所）长会议第三次会议联合公报

2006 年 5 月 17 日　韩国济州岛

5 月 16 日举办了以"东北亚国家空气质量管理"为主题的联合研讨会。该主题于 2005 年 9 月召开的 TPM3 工作层会议上被确定为优先考虑的环境问题。

院（所）长们进一步指出于 2005 年 9 月在韩国进行的工作层会议有助于 TPM3 的成功举行，并一致同意应当延续召开工作层会议的惯例，以助力 TPM4 的合作项目实施。

院（所）长们就各自院所为更好地适应环境价值观改变带来的挑战而新实行的改革进行了交流。他们对韩国国立环境科学院为鼓励对环境问题的全面解决而进行的重组改革表示认可。同时，院（所）长们特别关注日本国立环境研究所的第二个五年计划以及中国环境科学研究院的"十一五"计划。

在信息交流方面，院（所）长们同意指定各自院所的国际合作负责人作为联络人负责信息交流。英文版年度报告、重大研究项目清单和主要内容将存放于三国环境科研院（所）的图书馆以促进研究合作。此外，院（所）长们同意将 TPM 进展情况置于各院所的网站以便公众关注。

回顾 TPM2 达成一致的六个联合研究项目，即"淡水污染""大气污染，包括车源污染""跨界大气污染""黄沙尘暴""有害物质污染，如内分泌干扰物质及持久有机污染物"和"候鸟与湿地"时，院（所）长们一致同意由指定的联络人员负责共同实施上述项目。报告初稿（包括财务计划）经商讨后，将以合作研究项目提案的形式提交至下一次工作层会进行审查，并在 TPM4 进行进一步商讨。

院（所）长们同意在未来的 TPM 会上邀请三国的同类研究机构作为观察员。机构候选名单将提交至下一次的工作层会。作为观察员参会的机构应自行承担参会费用。此外，TPM4 应寻求纳入东北亚其他国家的同类科研机构，如朝鲜、蒙古国和俄罗斯。日韩两方院（所）长对中国环境科学研究院引入一家朝鲜环境研究机构成为 TPM 观察员所作出的努力表示赞赏。

院（所）长们回顾了于 2005 年 10 月 22 日至 23 日召开的第七次中日韩三国环境部长会议（TEMM）的成果，会上，部长们均对 TPM3 表示了支持。院（所）长们还对由三国环境科研院（所）共同参与的 TEMM 项目之一"淡水污染防治项目"一期工作的丰硕成果表示了赞赏。院（所）长们指出"东北亚空气污染物长距离越界传输"（LTP）项目改善了东北亚地区越界污染问题，并指出于 2005 年 11 月在韩国济州岛举行的 LTP 项目第八次专家会议十分成功。

院（所）长们进一步指出，2006 年春的黄沙尘暴问题非常严重。因此，他们一致同意全力支持三国环境科研院（所）合作开展黄沙尘暴联合研究，项目推进会议拟由中国环境科学研究院于 2006 年秋举办。

院（所）长重申三国环境科研院（所）将继续共同努力，加强日本国立环境研究所、中国环境科学研究院和韩国国立环境科学院之间的现有合作、深化未来合作，并同意继续将每年一次的 TPM 会议机制作为定期交流的平台。中国环境科学研究院院长提议于 2007 年春在中国举办 TPM4 及其平行研讨会，平行研讨会的主题将于 2007 年 2 月举行的工作层会议确定。

The Third Tripartite Presidents Meeting among NIER, CRAES and NIES Joint Communiqué

May 17, 2006, Jeju, Korea

The meeting was preceded on May 16th by an international workshop on "Air Quality Management in Northeast Asian Countries," a topic identified at the working level meeting for the Third TPM in September 2005 as a priority environmental issue.

The presidents further observed that the working level meeting held in September 2005 in Korea was conducive to the success of the Third TPM and agreed that the practice should be continued to assist the Fourth TPM in implementing joint projects.

The presidents exchanged information on the recent reforms taken place in their institutes to meet the challenge of changes in environmental values. They recognized the new paradigm of reorganization at NIER which aims to encourage holistic solutions to environmental problems. The presidents also expressed special interest in the Second 5-year plan of NIES and the Eleventh 5-year plan of CRAES.

On the topic of information exchange, the presidents agreed to assign the officials in charge of international cooperation at each institute as focal points for exchanging information. English language copies of annual reports, and the lists and the gist of major research projects will be deposited in the institutes' libraries to facilitate research cooperation. In addition, the presidents agreed to post the progress of TPMs on each institute's website for public awareness.

Recalling the six joint research projects agreed in the Second TPM, on "freshwater pollution," "air pollution including vehicular sources," "transboundary air pollution," "yellow sand storm," "hazardous materials contamination such as endocrine disrupting chemicals and POPs," and "migratory birds and wetlands," the presidents agreed that the implementing scheme of cooperative research will be jointly developed by the responsible focal points. The draft report of the consultation, including financial plans, will be submitted and reviewed at the next working level meeting in the form of cooperative research proposals to be considered at the Fourth TPM.

The presidents further agreed to welcome representatives of kindred research institutes in the three countries as observers in future TPMs. Lists of such candidate research institutes will be submitted to the next working level meeting. It is recommended, however, that the observers bear their costs of participation. In addition, the Fourth TPM should seek to incorporate kindred research institutes of other countries in Northeast Asia: North Korea, Mongolia and Russia. The presidents of NIER and NIES applauded efforts of CRAES to engage a North Korean environmental research institute as a TPM observer.

The presidents reviewed the outputs of the Seventh Tripartite Environment Ministers

Meeting (TEMM) held on October 22–23, 2005, at which the Ministers expressed support for the Third TPM. The presidents also expressed gratification at the fruitful result of the 1st phase "Freshwater Pollution Prevention Project" that was carried out as one of the TEMM projects with the participation of NIER, CRAES and NIES. The presidents also acknowledged that the "Long–range Transboundary Air Pollutants in Northeast Asia" (LTP) project contributes to the improvement of transboundary pollution problems in Northeast Asia and that the 8th Expert Meeting for LTP held in November 2005 in Jeju was successful.

The presidents further recognized severe episodes of yellow sand storm in spring 2006. In response, they agreed to give full support for a joint project among the three institutes of "yellow sand storm." The meeting to develop the project will be planned by CRAES in the autumn of 2006.

The presidents reaffirmed that the three institutes should continue the joint efforts to strengthen the existing and evolving cooperation among them, and agreed to continue the annual TPM as a vehicle for regular communications. President of CRAES offered to host the Fourth TPM in China in the spring of 2007, including a workshop whose theme will be decided at the Working Level Meeting in China in February 2007.

中日韩三国环境科学研究院（所）长会议第四次会议联合公报

2007 年 5 月 16 日　中国成都

TPM4 平行研讨会的主题是"车源排放等大城市的空气污染"。应成都市环保局要求，平行研讨会面向当地的环境管理人员和技术人员开放，以助力改善当地城市空气质量。院（所）长们认为该举措是 TPM 开展外展活动的范例。在平行研讨会上，院（所）长们听取了来自中国环境科学研究院、日本国立环境研究所、韩国国立环境科学院、蒙古国国家气象水文环境监测局和成都市环境科学研究院关于城市空气污染、机动车排放及其对人体健康影响的学术报告。

院（所）长们强调，TPM4 作为 TPM 会议机制第二个轮回的开始，是在前三次会议所获经验的基础上召开的。会议确立了"合作、交流、友谊、共赢"的原则。院（所）长们指出 TPM4 在 TPM3 的基础上，对 TPM 机制和会议形式均作了改善，并指出 TPM4 有助于加强三国环境科研院（所）间的合作交流。

院（所）长们对由三国环境科研院（所）于 2006 年 12 月 3 日至 5 日在北京共同举办的"区域生态与环境效应国际研讨会——沙尘暴及其影响和减缓措施"表示满意，并十分赞赏中国环境科学研究院将沙尘暴研讨会促升为 TEMM8

的边会，尤其是中日韩三国环境部长出席研讨会开幕式，极大地提升了研讨会和TPM 会议的水平和知名度。院（所）长们认为该研讨会水平较高，有助于推进中日韩三国在这一领域的联合研究。

院（所）长们指出 TPM4 已经实现了 TPM3 所设立的目标，即成功邀请到蒙古国国家气象水文环境监测局和中国中日友好环境保护中心作为观察员参会。

院（所）长们重申三方院所应当继续加强信息、出版物以及专家间的交流。为此，韩国国立环境科学院院长提议邀请中国环境科学研究院的研究人员参加韩国国立环境科学院的短期研修项目，并邀请日本国立环境研究所参与授课。日本国立环境研究所所长表示将斟酌此提议。

同时，院（所）长们再次确认了在 TPM2 上确立，并在 TPM3 上强调的六个优先合作领域。

院（所）长们讨论了关于沙尘暴（DSS）的联合合作项目，同意先启动 DSS 联合项目，成立一个工作组，并指派其于 2007 年 9 月底前汇报联合项目的进展情况。日本国立环境研究所所长强调了研究审核和数据分享对推进联合项目的重要性。中国环境科学研究院院长和韩国国立环境科学院院长持相同观点。

院（所）长们还听取了蒙古国家气象水文环境监测局和中国中日友好环境保护中心代表对其各自机构和研究活动的介绍。

日本国立环境研究所所长提议于 2008 年秋在日本举办 TPM5 及其平行研讨会，平行研讨会的主题将于 2008 年夏在日本举行的工作层会议确定。

The Fourth Tripartite Presidents Meeting among CRAES, NIES and NIER Joint Communiqué

May 16, 2007, Chengdu, Sichuan, China

The theme of TPM4 parallel workshop was "Air Pollution in Big Cities including Vehicle Exhaust". In response to request of Chengdu Environmental Protection Bureau (EPB), the workshop was open to local environmental management officers and technical staff to help them in improving urban air quality. The presidents recognized that this arrangement was a good example of outreach activity of TPM. In the parallel workshop, the presidents listened to academic reports from CRAES, NIES, NIER, NAMHEM[1] and CDAES[2] on urban air pollution, vehicle emission and impact on human health.

The presidents stressed that as the commencement of the second samsara of TPM meetings, TPM4 was held on the basis of experience gained in preceding meetings. The principle of "Cooperation, Communication, Friendship and All–Win" was developed.

---

[1]　National Agency for Meteorology, Hydrology and Environment Monitoring of Mongolia.

[2]　Chengdu Academy of Environmental Sciences.

They observed TPM4's improvement of TPM mechanism and meeting form on the basis of TPM3. They noted contributions of TPM4 towards enhancing cooperation and communication among CRAES, NIES and NIER.

The presidents expressed their satisfaction with the "International Workshop on Regional Ecology and its Environmental Effect – Dust Sand Storm, its Impact and Mitigation Countermeasure" organized by CRAES, NIES and NIER in Beijing on December 3–5, 2006. It was appreciated greatly that CRAES promoted the workshop as the side event of TEMM8, especially that the attending of Environment Ministers of China, Japan and Korea to the opening ceremony had elevated the level and popularity of both the workshop and TPM. The presidents considered that the workshop was of high–level, being helpful for promoting joint studies in this field among China, Japan and Korea.

The presidents noted that TPM4 had achieved the target set up by TPM3, successfully inviting NAMHEM and SJFEPC[1] to attend the meeting as observers.

The presidents reaffirmed that the three institutes should continue to exchange information, publications and expertise. In this regard, President of NIER proposed inviting researchers of CRAES to the short–term training program of NIER, and also requested NIES to join for giving lectures. President of NIES mentioned to consider this proposal.

The presidents also reaffirmed the six priority research areas established in TPM2 and

---

[1] Sino-Japan Friendship Environmental Protection Center of China.

emphasized in TPM3.

The presidents discussed the joint cooperative project on dust sand storm (DSS) and agreed to start DSS joint project firstly. The presidents agreed to establish a working group and instructed it to report the progress made on development of the joint project by the end of September 2007. President of NIES underlined the importance of research reviewing and data sharing to forward the joint project. President of CRAES and President of NIER shared the view.

The presidents learned from NAMHEM and SJFEPC their organization and activities.

President of NIES offered to host the Fifth TPM in Japan in the fall of 2008 including a workshop whose theme will be decided at the Working Level Meeting in Japan in the summer of 2008.

中日韩三国环境科学研究院（所）长会议第五次会议联合公报

2008 年 11 月 26 日　日本北海道札幌市

　　院（所）长们重申了 TPM2 确立的六大优先合作领域。未来三方将在这些领域不断探索合作项目的可能性。他们一致认为"淡水污染"是一个重要的研究主题，并赞成针对该主题的联合研究制作一份单独的讨论记录。关于"大气污染，包括车源污染"主题，院（所）长们同意三方都应选择各国的一座城市作为城市空气质量模型对比研究的地点，并在联系交流过程中生成一份研究计划。院（所）长们对于"跨界大气污染"相关合作的稳步推进表示十分满意。他们对沙尘暴（DSS）工作组提交的有关"黄沙尘暴"的联合研究项目计划表示认可，并建议就项目初期实施进一步制订详细计划。院（所）长们对"有害物质环境污染国际研讨会"上发起的"有害物质污染"相关合作活动表示欢迎。关于"候鸟与湿地"主题，院（所）长们一致认为该主题下的生物多样性保护至关重要。

　　此外，院（所）长们一致同意增加两个优先合作领域：一是固体废物管理，包括风险管理和 3R 原则；二是气候变化，包括气候变化适应和气候变化对生态系统的影响。院（所）长们就上述两个新领域指定了联络人，以进一步缩小议题范围并探索开展合作研究项目的可能性。

院（所）长们一致认为人员交流将极大地加强三方间的联系且有助于提高研究能力。他们就各自的长期、短期人员交流计划交换了意见，并一致同意尽可能地通过这些计划接收或派遣研究人员。

The Fifth Tripartite Presidents Meeting among NIES, NIER and CRAES Joint Communiqué

November 26, 2008, Hokkaido, Japan

The three presidents reaffirmed the six priority research areas identified at the TPM2 in which possibilities for joint projects would be explored. They shared the view that "freshwater pollution" is an important subject and agreed to produce a separate Record of Discussion on cooperative research regarding this subject. As for "air pollution including vehicular sources", the presidents agreed that each institution should select one city in each country as a location for comparative research on models concerning urban air quality and that a research proposal would be developed through the points of contact. The presidents expressed satisfaction with the steady development made in the collaboration on "transboundary air pollution". They also acknowledged a Joint Cooperative Project Proposal on "yellow sand storm" developed by the Working Group on Dust and Sand Storms (DSS) and recommended further elaboration toward its early implementation. The presidents welcomed the launch of collaborative activity on "hazardous material contamination" by the "International Workshop on Environmental Contamination by Hazardous Substances". As for "migratory birds and wetlands", the presidents shared the view of the importance of biodiversity conservation under this subject.

Furthermore, the three presidents agreed to add two new priority research areas. One is solid waste management, including risk management and 3Rs, and the other is climate change, including adaptation to climate change and the impact of climate change on ecosystems. The presidents nominated the points of contact for these two new areas to narrow down the subject areas and to explore the possibility of collaborative research projects.

The three presidents shared the view that personnel exchange greatly contributes to strengthening the ties among the three institutions and enhancing their research capacity. They also exchanged information on their respective short–term and long–term personnel exchange programs and agreed to accept or dispatch researchers under these programs where possible.

中日韩三国环境科学研究院（所）长会议第六次会议联合公报

2009 年 11 月 27 日　韩国首尔

　　院（所）长们就各自院所的最新进展以及共同面临的全球挑战进行了交流，并赞成加强合作以应对气候变化、实现低碳社会、资源回收利用和废弃物管理等问题。

　　此外，三国环境科研院（所）回顾了在优先合作领域的活动进展。院（所）长们对"跨界大气污染"和"沙尘暴"（DSS）合作项目的稳步推进表示满意。日本国立环境研究所通过"日本—中国水环境伙伴关系"开展了淡水污染防治研究，并欢迎韩国国立环境科学院和中国环境科学研究院加入合作。韩国国立环境科学院邀请日本国立环境研究所和中国环境科学研究院参加拟于 2009 年 12 月在中国青岛举办的"持久性有机污染物论坛"。

　　韩国国立环境科学院提议了一个关于回收物风险评估及安全准则的新合作项目。中国环境科学研究院和日本国立环境研究所对该提议表示支持，并强调三方相关领域专家尽早就项目实施作进一步探讨。

　　院（所）长们一致认同人员交流将加强三国环境科研院（所）之间的研究合作，

促进东亚地区环境研究领域的信息共享。

院（所）长们同意将研讨会面向中日韩三国的大学和研究所的环境领域相关专家公开，并接受蒙古国和印度作为观察员参与 TPM7。中国环境科学研究院建议三方在 2010 年 10 月于日本举办的《生物多样性公约》第 10 次缔约方大会（COP10）上联合组织一个边会。日本国立环境研究所将联系日本环境省进一步讨论该提案的可行性。

The Sixth Tripartite Presidents Meeting among NIER, CRAES and NIES Joint Communiqué

November 27, 2009, Seoul, Korea

The three presidents exchanged information on the recent developments in each institute and shared the view of the global challenges we are facing. They agreed to strengthen cooperation to address the issues including climate change, realization of low carbon society, resource recycling and waste management.

Furthermore, the three institutes reviewed the progress of their activities in the priority research areas. The presidents expressed satisfaction with the steady development made in the collaboration on "Transboundary Air Pollution" and "Dust and Sand Storm" (DSS). Freshwater pollution prevention study has been carried out with Japan–China Water Environment Partnership and NIES welcomed the participation of NIER and CRAES to the partnership. NIER invited NIES and CRAES to attend the "POPs Forum" to be held in December 2009 in Qingdao, China.

NIER proposed a new cooperation project on Risk Assessment and Safety Guideline on Recycled Product. CRAES and NIES agreed on this proposal and emphasized that experts in this field from the three institutes should further discuss the implementation in detail at the earliest opportunity.

The three institutes acknowledged the importance of the exchange of experts for enhancing research collaboration and sharing information on environmental research in East Asia.

The three presidents agreed to open the workshop to environment related experts from universities and research institutes from the three countries, as well as accepting the participation of Mongolia and India as observers of TPM7. CRAES suggested that the three institutes jointly organize a side event at the COP10 of CBD in October 2010 in Japan. NIES will approach the Ministry of the Environment, Japan, to discuss the possibility of this proposal.

中日韩三国环境科学研究院（所）长会议第七次会议
联合公报

2010 年 9 月 14 日　中国青岛

　　TPM7 平行研讨会的主题是"生物多样性保护与固体废物管理"。为向更多受众广泛介绍TPM会议机制，此次研讨会特向当地环境管理官员和研究人员开放。院（所）长们认同三国环境科研院（所）与相关感兴趣的组织共享专业知识有切实益处。会上，院（所）长们听取了来自中国环境科学研究院、日本国立环境研究所、韩国国立环境科学院、青岛市环境科学研究院等关于各个相关议题的报告，并重申了这些研究领域的重要性。

　　院（所）长们强调，作为 TPM 会议机制第三轮的开始，TPM7 是在往届会议所积累经验的基础上召开的。他们指出了 TPM 对三国环境科研院（所）间进一步合作交流所作出的贡献，并认为 TPM7 是在往年基础上对 TPM 机制的进一步发展。院（所）长们对联合国环境规划署代表的参与表示了欢迎，希望 TPM 在联合国环境规划署的支持下，将能够在更广泛的区域合作中起到积极作用。

　　院（所）长重申三方应继续进行信息、出版物以及专业知识的交流。他们同意进一步加强八大优先合作领域联络人之间的联系。三方分别就淡水污染防治、跨界空气污染、汞监测、气候变化与固体废物管理合作项目提出了建议。院（所）长们同意将优先合作领域中的"候鸟与湿地"更名为"生物多样性保护"，以扩大今后合作的潜力。院（所）长们还讨论了即将于 2010 年 10 月在日本名古屋召开的生物多样性公约（CBD）缔约方大会第 10 次会议（COP10）边会的组织情况，各院所开展的活动将对另外两方发出邀请。

The Seventh Tripartite Presidents Meeting among CRAES, NIES and NIER Joint Communiqué

September 14, 2010, Qingdao, Shandong, China

The theme of TPM7 parallel workshop was "Biodiversity Conservation and Solid Waste Management". In order to introduce TPM to a wider audience, the workshop was opened to local environmental management officials and researchers. The presidents recognized the practical advantages of allowing the three institutes to share their expertise with related interest groups. The presidents listened to presentations from CRAES, NIES, NIER, and QDRAES[1] on various related topics and reaffirmed the importance of these research areas.

The presidents stressed that as the third round of the Tripartite Presidents Meeting commences, TPM7 was held on the basis of the experience accumulated from previous meetings. They noted the contributions of TPM7 to further cooperation and communication among CRAES, NIES and NIER and acknowledged TPM7's

---

[1] Qingdao Research Academy of Environmental Sciences.

development of the TPM mechanism from previous years. The Presidents welcomed the attendance of the UNEP representatives and envisioned that TPM could play an active role in broader regional collaboration supported by UNEP.

The presidents reaffirmed that the three institutes should continue to exchange information, publications and expertise. They agreed to further enhance the contact among focal points of the eight priority research areas. Each institute made proposals for cooperative projects, including fresh water pollution prevention, transboundary air pollution, mercury monitoring, climate change and solid waste management. The three presidents agreed to rename the priority research area of "migratory birds and wetlands" as "biodiversity conservation" to expand the potential for future collaboration. They also discussed their organization of side events at the upcoming 10th meeting of the Conference of the Parties (COP10) to the Convention on Biological Diversity (CBD) to be held in October in Nagoya, Japan, and offered invitations to each other's respective events.

中日韩三国环境科学研究院（所）长会议第八次会议联合公报

2011 年 11 月 22 日 日本冲绳岛

院（所）长们强调 TPM8 是在往届会议积累的经验基础上召开的，TPM8 对加强今后三方之间的合作、沟通、信息及人员交流具有重要贡献。此外，院（所）长们肯定了各自院所对于巩固和发展 TPM 机制所作出的贡献。

院（所）长们同意对 TPM 八大优先合作领域的名称、定义及内容作部分修订，以符合当今世界不断变化的环境状况。经商讨，优先合作的领域更改为，淡水污染、亚洲大气污染、城市环境和生态城市、沙尘暴、化学品风险及管理、生物多样性保护、固废管理和气候变化。

院（所）长们承诺进一步加强三方在上述研究领域工作组成员和联络员间的联系，以推进三国环境科研院（所）的合作项目。院（所）长们一致同意邀请 TPM 机制外其他国家中对 TPM 八大优先合作领域感兴趣的相关研究人员和机构参与今后的 TPM 会议。

TPM8 平行研讨会的主题是"亚洲大气污染和生物多样性保护"。为了更好地传播 TPM 目标和成果，研讨会向地方环保官员、研究员和其他感兴趣的旁听者开放。研讨会广泛覆盖 TPM 优先合作领域，并再次强调了三方在这些领域合作的重要性。

院（所）长们一致认同三方与相关机构和组织分享专业知识的重要性。院长（所）们认真聆听了来自日本国立环境研究所、韩国国立环境科学院、中国环境科学研究院、那霸地区自然保护办事处（日本环境省）、冲绳县卫生环境研究所和琉球大学代表及研究员的报告。

The Eighth Tripartite Presidents Meeting among NIES, NIER and CRAES Joint Communiqué

November 22, 2011, Okinawa, Japan

The presidents stressed that TPM8 was held on the basis of the experience accumulated from previous meetings, as well as the important contribution of TPM8 to furthering cooperation, communication, and information and personnel exchange among NIES, NIER and CREAS. In addition, the presidents acknowledged the respective institution's contributions to nurturing and consolidating the TPM mechanism at TPM8.

The presidents agreed to partially refine the names, definitions and content of the eight TPM Priority Research Areas to reflect the changing environmental circumstances in the world. The names of the Priority Research Areas were agreed as follows – Freshwater Pollution; Asian Air Pollution; Urban Environment and Eco–city; Dust and Sand Storm; Chemical Risk and Management; Biodiversity Conservation; Solid Waste Management; and Climate Change.

The presidents committed to further enhancing the contact among working group members and focal points in these research areas in order to advance cooperative projects among the three institutions. The presidents agreed to put forward the idea of extending invitations to researchers and parties on the eight TPM Priority Research Areas from concerned countries falling outside the TPM mechanism to attend future TPM meetings.

The theme of the TPM8 Parallel Workshop was "Asian Air Pollution and Biodiversity Conservation". The workshop was opened to local environmental officials, local researchers and other interested observers, in order to facilitate the dissemination of the TPM goals and outcomes. The workshops, encompassing the broad scope of the TPM Priority Research Areas reaffirmed the importance of extant and forthcoming collaborations among the three institutions in these areas.

The presidents acknowledged the importance and practical advantages of the three institutions sharing their accumulated expertise with related institutions and interested parties. Presentations by delegates and researchers from NIES, NIER, CRAES, Naha Regional Office for Nature Conservation (Ministry of the Environment Japan), Okinawa Prefectural Institute of Health and Environment, and University of the Ryukyus were received with interest by the presidents.

中日韩三国环境科学研究院（所）长会议第九次会议联合公报

2012 年 11 月 13 日　韩国平昌

　　院（所）长们强调，TPM9 是在往届会议所积累的经验基础上召开的，并强调 TPM 在进一步推动三方合作以及通过人员信息交流和三国环境科研院（所）间的交流加强相互协作方面的重要性。此外，三国环境科研院（所）就各自最近的进展以及研究活动进行了交流，并对各自院所在 TPM9 期间为巩固 TPM 机制所作出的贡献表示认可。

　　院（所）长们重申三国环境科研院（所）应继续加强合作，进一步加强信息、数据、出版物及专业知识的互联互通。他们同意促进八大优先合作领域的联络员间的联系往来，并肯定了在每个优先合作领域内提出具体措施以促进合作的重要性。院（所）长们同意在每个优先合作领域选定一个牵头协调机构。牵头协调机构的主要职责应包括信息交流、组织培训、举办会议、研讨会及专题研讨会、开展新合作项目。院（所）长们决定，原则上由一个牵头协调机构对每个优先合作领域负责一年，但在三方同意的情况下负责期限可以延长。各优先合作领域的牵头协调机构将在 TPM10 上展示各研究领域的成果。

经讨论，各优先合作领域的牵头协调机构安排如下：

韩国国立环境科学院——沙尘暴、固体废物管理

中国环境科学研究院——淡水污染、化学品风险与管理、生物多样性保护

日本国立环境研究所——亚洲大气污染、城市环境与生态城市、气候变化

TPM9 平行研讨会的主题是"城市环境和生态城市，气候变化的影响与适应"。来自三国环境科研院（所）的参会者们分享了研究成果。院（所）长们表达了在这两个优先合作领域进一步深化合作的意愿。

The Ninth Tripartite Presidents Meeting among NIER, CRAES and NIES Joint Communiqué

November 13, 2012, Pyeongchang, Korea

The presidents stressed that TPM9 was held on the basis of the experience accumulated from previous meetings and the important role of TPM9 in furthering cooperative endeavors, and strengthening mutual collaboration by means of personnel and information exchange and discussions among NIER, CRAES and NIES, as facilitated by the efforts of the Focal Points for the eight Priority Research Areas (PRAs). In addition, the three institutes exchanged information on their recent developments and research activities and acknowledged respective institution's contributions to consolidating the TPM mechanism at TPM9.

The presidents reaffirmed that the three institutes should continue to strengthen collaborations based on the exchange of information, data, publications and expertise. They agreed to further enhance the contact among Focal Points of the eight PRAs and also recognized the importance of establishing concrete measures to further facilitate cooperation in each PRA. The presidents agreed to select a lead coordinating institute for each PRA. Representative responsibilities of a lead coordinating institute should include information exchange, organization of training, workshop, seminar and symposium and development of new cooperative projects. The presidents decided in principle that a lead coordinating institute is in charge of each PRA for one year, but the term could be extended upon mutual agreement and the lead coordinating institute for each PRA would present the outcomes of each research item at TPM10.

Based on discussions, lead coordinating capacity for each PRA was assigned as follows:
NIER   – Dust and Sand Storm, Solid Waste Management
CRAES – Freshwater Pollution, Chemical Risk and Management, Biodiversity Conservation
NIES   – Asian Air Pollution, Urban Environment and Eco–city, Climate Change

The theme of the TPM9 Parallel Workshop was "Urban Environment and Eco–city, and Climate Change Impact and Adaptation". NIER, CRAES and NIES shared their research findings with the participants and the presidents expressed their willingness to engage in further collaborations in these two PRAs.

中日韩三国环境科学研究院（所）长会议第十次会议联合公报

2013 年 11 月 6 日　中国南京

　　院（所）长们听取了三国环境科研院（所）关于自 TPM9 以来各自院所研究和发展活动的报告，以及有关灾害性环境问题的研究和之前八个优先合作领域的进展情况。院（所）长们对三国环境科研院（所）为促进 TPM 合作发展作出的共同努力表示满意。

　　院（所）长们重申三国环境科研院（所）应继续交流信息、出版物和专业知识。院（所）长们一致认为，应在考虑三国环境科研院（所）系统性合作的情况下开展优先合作领域的合作，每个优先合作领域的联络员应更积极地参与 TPM 活动。

　　"灾害性环境问题"被确立为新的优先合作领域。经讨论，在 TPM11 会议前，各优先合作领域的牵头协调机构重新分配如下：

　　中国环境科学研究院——淡水污染、城市环境与生态城市、化学品风险与管理

　　日本国立环境研究所——生物多样性保护、气候变化、灾害性环境问题

　　韩国国立环境科学院——亚洲大气污染、沙尘暴、固废管理

　　TPM10 平行研讨会的主题是"淡水污染"。为扩大 TPM 会议的影响，研讨

会与"国际清洁水行动环境保护研讨会"归并举行。该研讨会由江苏省环保局主办，由江苏省环境科学研究院、中国环境科学研究院、韩国国立环境科学院和日本国立环境研究所协办。平行研讨会向来自中国、日本、韩国和其他国家（如澳大利亚和荷兰）的环境管理官员和研究人员开放。院（所）长们对与国际专家进行经验交流分享的机会表示感谢。

院（所）长们强调，随着 TPM 会议机制第四轮的开始，TPM10 将继续秉持"友谊、交流、合作、共赢"的基本原则。院（所）长们表示 TPM10 有助于促进三国环境科研院（所）间的进一步合作交流，并赞同 TPM10 推动了优先合作领域的联合研究。院（所）长们一致认为，在新环境形势下，TPM 将在东北亚乃至全球环境改善中发挥越发重要的作用。

The Tenth Tripartite Presidents Meeting among CRAES, NIES and NIER Joint Communiqué

November 6, 2013, Nanjing, China

The Presidents listened to the reports from the three institutes on their research and development activities since TPM9, researches related to disaster environment and the progress on the previous eight priority research areas (PRAs). The Presidents noted the joint efforts made by the three institutes in the development of TPM cooperation and expressed their satisfactions.

The Presidents reaffirmed that the three institutes should continue to exchange information, publications and expertise. The Presidents agreed that the cooperation on PRAs should be carried out with consideration of systematic collaboration among CRAES, NIES and NIER and the focal points of each PRA should be involved more in TPM activities.

"Disaster Environment" was identified as a new PRA. Based on the discussions, the lead coordinating institute for each PRA until TPM11 was reallocated as follows:

CRAES – Freshwater Pollution, Urban Environment and Eco–city, Chemical Risk and
           Management

NIES   – Biodiversity Conservation, Climate Change, Disaster Environment

NIER   – Asian Air Pollution, Dust and Sand Storm, Solid Waste Management

The theme of TPM10 Parallel Workshop was "Freshwater Pollution". In order to broaden the influence of TPM, the workshop was integrated with the International Environmental Protection Symposium on Clean Water Action, which was organized by Jiangsu EPB and co–organized by JSAES[1], CRAES, NIES and NIER. The Parallel Workshop was opened to environmental management officials and researchers of China, Japan, Korea and other countries, such as Australia and the Netherlands. The Presidents appreciated the opportunity of sharing experiences with international experts.

The Presidents stressed that as the fourth round of TPM commences, TPM10 was held in adherence to the basic principle of "Friendship, Communication, Cooperation and Win–Win". The Presidents noted the contributions of TPM10 to further cooperation and communication among CRAES, NIES and NIER and acknowledged TPM10's further promotion of joint study on PRAs. The Presidents reached the common recognition that under the new environmental situation, TPM will play a more and more important role in the environmental improvement in Northeast Asia and the world.

---

[1] Jiangsu Provincial Academy of Environmental Science.

中日韩三国环境科学研究院（所）长会议第十一次会议联合公报

2014 年 11 月 13 日　日本川崎

院（所）长们均强调，三方的务实合作对改善东北亚地区环境至关重要。

院（所）长们分享了各自院所的最新进展情况，以及"$PM_{2.5}$ 和短寿命气候污染物"专题研究进展，该主题被确立为三国亟待解决的问题。院（所）长们一致认为，这是三方在大气污染问题上合作的第一步，希望继续将信息交流与合作作为未来合作的重点。

院（所）长们听取了各牵头协调机构联络员关于优先合作领域的报告。经讨论决定，各优先合作领域的牵头协调机构在 TPM12 会议前保持如下：

日本国立环境研究所——生物多样性保护、气候变化、灾害性环境问题

韩国国立环境科学院——亚洲大气污染、沙尘暴、固体废物管理

中国环境科学研究院——淡水污染、城市环境与生态城市、化学品风险与管理

院（所）长们一致认为，在优先合作领域开展合作至关重要，各优先合作领域牵头协调机构联络员应在 TPM 机制下引领三方合作，共享各优先合作领域的信息，以便及时推动相关交流。

为扩大 TPM 机制下的活动，院（所）长们一致赞同寻求与中日韩其他机构的特定研究人员开展合作。日本国立环境研究所所长还介绍了与国际组织合作的可能性。院（所）长们均同意就此作进一步讨论。

TPM11 平行研讨会的主题是"生态城市和生物多样性"。院（所）长们指出，这两个主题是中日韩三国重要的环境问题。院（所）长们对分享三方研究成果的相关信息和进展表示非常满意，并强调深化合作的重要性。院（所）长们对来自日本国立环境研究所、韩国国立环境科学院、中国环境科学研究院、日本环境省和川崎市的所有参会人员表示感谢。

The Eleventh Tripartite Presidents Meeting among NIES, NIER and CRAES Joint Communiqué

November 13, 2014, Kawasaki City, Japan

The Presidents stressed the importance of the practical and pragmatic joint efforts among the three institutes to improve the regional environment in Northeast Asia.

The Presidents shared recent development in the three institutes' activities and special research topics related to $PM_{2.5}$ and SLCP (Short–Lived Climate Pollutant) identified as urgent issues in the three countries. They agreed that this is the first step for collaboration among the three institutes on the issue of atmospheric pollution, and hope that information exchange and collaboration are continued as a matter of high priority into the future.

The Presidents listened to the reports on each Priority Research Area (PRA) from the Focal Points (FPs) of the respective Lead Coordinating Institutes (LCIs). Based on the discussions, it was decided that the LCI for each PRA until TPM12 will remain as follows:

NIES   – Biodiversity Conservation, Climate Change, Disaster Environment

NIER   – Asian Air Pollution, Dust and Sand Storm, Solid Waste Management

CRAES – Freshwater Pollution, Urban Environment and Eco–city, Chemical Risk and
          Management

The Presidents agreed that cooperation in the PRAs is of utmost importance, and that the FPs of the LCI in each PRA should lead the collaboration among NIES, NIER and CRAES under the TPM mechanism; and that FPs for the WG should share information on each PRA to facilitate relevant communication in a timely manner.

To expand TPM activities, the Presidents agreed to seek the collaboration of selected researchers in other institutions of the three countries. President of NIES also introduced the potential for collaboration with international organizations. The Presidents agreed to follow up this issue with further discussions.

The theme of TPM11 International Workshop was "Eco–city and Biodiversity". The Presidents noted that both themes are important environmental issues in Japan, Korea, and China. They are satisfied with the sharing of information on recent situations and research findings in the three countries, and emphasized the importance of further collaboration.

中日韩三国环境科学研究院（所）长会议第十二次会议联合公报

2015 年 11 月 4 日　韩国丽水

院（所）长们分享了三国环境科研院（所）最近的研究活动以及"气候变化"专题研究进展，并一致认为气候变化是三国亟待解决的问题。院（所）长们听取了各牵头协调机构联络员关于各优先合作领域的报告。经讨论决定，各优先合作领域的牵头协调机构在 TPM13 前保持如下：

韩国国立环境科学院——亚洲大气污染

　　　　　　　　　沙尘暴

　　　　　　　　　固体废物管理

中国环境科学研究院——淡水污染

　　　　　　　　　城市环境与生态城市

　　　　　　　　　化学品风险与管理

日本国立环境研究所——生物多样性保护

　　　　　　　　　气候变化

　　　　　　　　　灾害性环境问题

院（所）长们一致认为，优先合作领域的积极有效合作是延续三方成功合作的关键。院（所）长们审阅了 2015—2019 年路线图草案，概述了优先合作领域的发展愿景。院（所）长们原则上同意路线图，但认为可进一步修改。基于路线图的三方合作预计将有助于改善亚洲人民的生活。

院（所）长们强调，TPM 机制应继续秉持"友谊，交流，合作，共赢"的原则。TPM12 平行研讨会的主题是"亚洲空气污染"。三国环境科研院（所）交流了研究成果，分别介绍了各自关于减缓空气污染的策略或政策，并介绍了空气质量和预报系统的现状。此外，来自美国国家航空航天局（NASA）和法国冰川和环境地球物理实验室（LGGE）的专家也参加了研讨会，就亚洲的空气污染问题进行了报告，与会专家对于未来的研究方向进行了讨论。三位院（所）长希望通过积极讨论三国的问题、经验、技术和措施，以进一步加强亚洲空气污染联合研究。

The Twelfth Tripartite Presidents Meeting among NIER, CRAES and NIES Joint Communiqué

November 4, 2015, Yeosu, Korea

The presidents shared recent developments in the three institutes' activities and special research topics related to "Climate Change", identified as urgent issues in the three countries. The presidents listened to the reports on each PRA from the Focal Points of the respective Lead Coordinating Institutes (LCIs). Based on the discussions, it has been decided that the LCI for each PRA until TPM13 will remain as follows:

NIER – Asian Air Pollution
Dust and Sand Storm
Solid Waste Management

CRAES – Freshwater Pollution
Urban Environment and Eco–city
Chemical Risk and Management

NIES – Biodiversity Conservation
Climate Change

Disaster Environment

The three presidents agreed that active and effective cooperation in the PRAs is key to the ongoing success of collaborations. The presidents reviewed the draft of the Roadmap for the period 2015–2019, outlining a vision for the PRAs development. Also the presidents agreed in principle to the Roadmap which however, can be modified. The tripartite cooperation based on the Roadmap is expected to contribute to the improvement of people's life in Asia.

The presidents emphasized that the TPM should continue to be developed under the principle of "Friendship, Exchange, Cooperation and Win–Win." The theme of TPM12 International Workshop was "Asian Air Pollution." The three institutes exchanged the research results, introducing their strategies or policies on air pollution reduction and presenting the current status of air quality and forecasting systems. Also, National Aeronautics and Space Administration (NASA), the USA and Laboratoire de Glaciologie et Geophysique de I'Environnement (LGGE), France participated in the workshop making presentations on air pollution in Asia as well as future research directions were discussed at the panel discussion. The presidents conveyed their willingness to further enhance the joint research on air pollution in Asia through active discussions on the three countries' problems, experiences, know–how and measures.

中日韩三国环境科学研究院（所）长会议第十三次会议
联合公报

2016 年 11 月 2 日　中国昆明

一、序言

1. 中国环境科学研究院、日本国立环境研究所和韩国国立环境科学院三方之间的中日韩三国环境科学研究院（所）长会议（TPM）机制由三方于 2004 年共同建立，旨在加强三方在环境保护领域发展和科研合作的信息和经验交流，以共同改善东北亚地区的环境质量。

2. 2016 年 10 月 31 日至 11 月 4 日，由中国环境科学研究院主办的第十三届中日韩三国环境科学研究院（所）长会议（TPM13）在中国昆明召开。应中国环境科学研究院副院长宋永会代表新任院长李海生发出的邀请，韩国国立环境科学院院长朴辰远和日本国立环境研究所所长住明正分别率团出席 TPM13。

二、中日韩环境事务和科学技术最新进展

3. 在会议开幕式上，院（所）长们就 TPM12 后各国环境事务和科学技术进展情况进行了交流。

4. 中国环境科学研究院副院长宋永会对来自日本和韩国的朋友表示热烈欢迎，

并充分肯定了三国环境科研院（所）利用 TPM 机制在优先合作领域进行的交流。宋永会副院长认为，TPM13 作为 TPM 机制新一轮的开端，需要解决四大议题：回顾 TPM 历史、总结成果、评估现状和展望未来。他强调，在未来的 TPM 机制建设中，三国环境科研院（所）应：（1）继续拓展、整合科研合作领域；（2）加强在环境基准和标准领域的深入合作；（3）在加强三国科研人员技术交流和沟通的同时，放眼亚洲，明确与全球环境相关的"热点问题"；（4）共同努力提高科研质量，最终成为环境研究和保护领域的全球领导者。宋永会副院长还提议，在关注优先合作领域的同时，需与科技企业和行业合作以进一步深化三方交流和创新。

5. 住明正所长在开幕致辞中简要介绍了日本国立环境研究所于 2016 年 4 月开始实行的第四个"中长期计划"。他强调了"网络、整合、发展、综合"四大关键词，诠释了日本国立环境研究所未来五年的战略，包括温室气体观测卫星（GOSAT）项目、日本环境与儿童研究研究（JECS）项目，以及新成立的气候变化战略合作办公室、促进社会对话和合作生产办公室。住明正所长简要介绍了日本国立环境研究所于 2016 年 4 月开设的福岛分所将开展灾害性环境问题研究项目，并介绍了关于拓展与当地研究机构合作活动的计划。

6. 朴辰远院长介绍了韩国国立环境科学院 2015 年的工作情况及本年度研究目标：营造安全舒适的环境，促进环境保护与经济发展共存，实现社会可持续发展。通过介绍韩国和东北亚地区的环境问题，朴辰远院长强调了 TPM 成员国在东北亚地区的重要作用和责任。针对最近黄沙和颗粒物引起的对空气污染加剧的担忧，朴辰远院长呼吁三国环境科研院（所）加强三边环境科研合作。在 TPM 优先领域合作路线图完成之际，朴辰远院长希望并相信路线图不仅有助于推进三国环境科研院（所）未来的联合研究项目，还将有助于后代解决全球悬而未决的环境问题。

三、研究进展和优先研究领域

7.院（所）长们听取了TPM12后的科研进展、环境健康和风险评估专题报告，以及9个优先合作领域的合作成果汇报。

8.院（所）长们还就更新九个优先合作领域交换了意见。韩国国立环境科学院表示，其"生物多样性保护"研究任务已于2013年转至韩国国家生态研究所（NIE），无法继续围绕该领域开展合作研究，并建议将该研究领域从优先合作领域清单中移除。此外，韩国国立环境科学院还建议鉴于人们对"有害藻华"的研究兴趣与日俱增，可将"有害藻华"列入TPM机制下的优先合作领域。对此，院（所）长们认为现有优先合作领域是三方共同认同的需优先解决的重大环境问题。院（所）长们一致同意将"生物多样性保护"保留为优先合作领域，并重申可邀请其他机构的研究人员参与相关研究。院（所）长们均表示优先合作领域为大研究领域，而"有害藻华"属于研究课题，可将其设为"淡水污染"的次级优先合作领域。

9.院（所）长们高度评价了近年来各院所取得的进展，并强调优先合作领域应基于各国环境研究优先次序。同时，区域性和全球性环境问题也应纳入考量，未来应进一步深化各优先合作领域的科学技术交流。

四、采纳《TPM优先领域合作路线图（2015—2019）》并讨论TPM机制未来发展

10.院（所）长们审议并通过了《TPM优先领域合作路线图（2015—2019）》修订版，并强调在此期间路线图也可进行修订。

11. 院（所）长们确认当前各优先合作领域的牵头协调机构保持如下：中国环境科学研究院（淡水污染、城市环境与生态城市、化学品风险与管理）；日本国立环境研究所（生物多样性保护、气候变化、灾害性环境问题）；韩国国立环境科学院（亚洲大气污染、沙尘暴、固废管理）。

12. 住明正所长指出环境问题范围宽泛，环境领域的研究兴趣和需求也多种多样。他强调，三国环境科研院（所）各有使命，情况也各不相同，并指出应具体到个人层面开展联合研究。住明正所长提议，应秉持 TPM 机制"友谊、交流、合作、共赢"的基本原则，共同讨论关于当前 TPM 机制形式的改革和推动优先合作领域产出具体成果的方法。

13. 院（所）长们指出，TPM 机制对提升区域环境保护技术具有重要意义，并强调三国环境科研院（所）应根据各国国情，加强合作，实现科学研究和环境技术发展"共赢"。院（所）长们还指出，各院（所）应利用各自能力优势，解决共同关心的环境问题。

五、TPM13 国际研讨会暨云南省环境科学研究院建院 40 周年国际学术交流会：水污染防治技术及生态系统健康

14. TPM13 国际研讨会暨云南省环境科学研究院建院 40 周年国际学术交流会于 2016 年 11 月 2 日召开。会议由 TPM 成员方和云南省环境科学研究院共同组织，会议主题是"水污染防治技术及生态系统健康"，旨在促进环境研究和科研成果的传播。云南省环境保护厅和云南省环境科学研究院的高层领导与专家出席会议并作了主旨报告。

15. 会上交流讨论了最新科研成果，包括流域藻华和水污染控制、流域水环境管理与生态恢复，以及水质标准和风险评估。

16.美国华盛顿大学环境与职业健康科学系斯韦勒·威德尔教授和蒙古国国立气象水文环境研究所所长甘珠尔·萨仁图雅应特别邀请参加了研讨会，并就"水污染与健康"作了专题报告。

六、TPM14

17.住明正所长提议于 2017 年在日本仙台市南东北地区举办 TPM14。

七、总结

18.院（所）长们对 TPM13 所取得的丰硕成果表示满意，并强调应在加强开拓优先合作领域联合研究的原则上，积极推动 TPM 机制，使三方环科院（所）的科技交流与合作更上一层楼，在改善东北亚区域环境质量和促进可持续发展方面发挥更重要的作用。

19.住明正所长和朴辰远院长对中国环境科学研究院、云南省环境保护厅和云南省环境科学研究院为成功举办 TPM13 所付出的努力及其热情款待表示衷心的感谢。

The Thirteenth Tripartite Presidents Meeting among CRAES, NIES and NIER Joint Communiqué

November 2, 2016, Kunming, China

I. Preface

1. The Tripartite Presidents Meeting of the Chinese Research Academy of Environmental Sciences (CRAES) of China, the National Institute for Environmental Studies (NIES) of Japan and the National Institute of Environmental Research (NIER) of Korea (TPM) was jointly established in 2004. TPM aims to enhance exchange of information and experiences of each institute's development and research collaboration in the field of environmental protection for better environment quality in Northeast Asia.

2. The 13th TPM (TPM13), organized by CRAES, was held in Kunming, China from October 31 to November 4, 2016. Delegations respectively headed by President SUMI Akimasa of NIES and President Park Jin–won of NIER attended TPM13 at the invitation of Vice President SONG Yonghui of CRAES on behalf of Dr. LI Haisheng who was appointed as the new President of CRAES.

II. The Latest Developments in Environmental Affairs and Science and Technology in the Three Countries

3. In the opening session, the Presidents exchanged information about the developments in environmental affairs and science and technology in their respective countries after TPM12.

4. Vice President SONG Yonghui of CRAES warmly welcomed the friends from Japan and Korea and fully affirmed the exchanges conducted in the priority research areas (PRAs) by the three institutes utilizing the TPM mechanism. With the beginning of this new round, TPM13, he identified four general topics to be addressed: review of the TPM history, summary of achievements, assessment of current status and identification of future ventures. He emphasized that in the future development of the TPM mechanism, the three institutes should: (1) continue to extend and integrate the areas of research collaboration; (2) strengthen the in–depth cooperation in the areas of environment criteria and standards; (3) take a broad view of Asia and identify the "hot issues" relating the global environment in addition to enhancing the technical exchanges and communication among the scientific researchers of the three countries; and (4) work together to improve the quality of the research to eventually become global leaders in the area of environmental research and protection. Vice President SONG Yonghui also proposed that while focusing on PRAs, we need to cooperate with technology enterprises and industry to even further tripartite exchange and innovation.

5. President SUMI Akimasa of NIES in his opening address introduced the overview of the Forth Medium–and–Long–Term Plan of NIES which began April 2016. He underlined the four keywords, "Network, Integrate, Evolve, Synthesize" which express the NIES's strategies for the next five years by noting some research projects such as

a greenhouse gases observing satellite (GOSAT) project and the Japan Environment and Children's Study (JECS), and a newly established Climate Change Strategy Collaboration Office and Office for Facilitating Social Dialogue and Co-production. President Sumi outlined the NIES Fukushima Branch inaugurated in April 2016 where the disaster environment research program would be undertaken. He also introduced a plan to extend the collaborative activities with local research institutes.

6. President PARK Jin-Won introduced the NIER's performance in 2015 and research goals this year: creating a safe and pleasant environment, promoting environmental and economic co-existence, and realizing a sustainable society. Presenting the environmental issues in Korea as well as Northeast Asia, he emphasized the significance of the role and responsibility of the TPM member countries in the region. Mentioning the recent concerns over worsening air pollution caused by yellow dust and particulate matter, President PARK encouraged the three institutes to further strengthen the trilateral cooperation in environmental research. Moreover, celebrating the completion of the TPM PRA Roadmap, he expressed his hope and belief that the Roadmap would contribute to not only the improvement of future joint research projects among the three institutes but also to solving pending environmental issues in the world for future generations.

III. Research Progress and the Priority Research Areas (PRAs)

7. The Presidents listened to the scientific research progress since TPM12, special reports on environmental health and risk assessment and collaboration results of the nine PRAs.

8. The Presidents also exchanged ideas for the renewal of the nine PRAs. NIER

explained that as its scientific research tasks for "Biodiversity Conservation" were transferred to the Korean National Institute for Ecology (NIE) in 2013, it has no capability to continue the collaborative research in this area, and suggested removing this research area from the priority list. In addition, NIER proposed adding "Harmful Algal Blooms" as a new priority area under the TPM mechanism because of the increasing interest in this issue. In response to this, the Presidents confirmed that the existing PRAs are the research areas which they identified as the major environmental issues to be addressed on a priority basis. They agreed not to remove "Biodiversity Conservation" and reconfirmed that researchers belonging to other institutes can be invited. The Presidents shared the view that the PRAs are large research areas while "Harmful Algal Blooms" is a research topic which could be a sub–PRA of "Freshwater Pollution".

9. The Presidents spoke highly of the progress made among the three institutes in recent years, and emphasized that the PRAs should be based on environmental research priorities in each country. Meanwhile, regional and global environmental issues should be taken into account, and scientific and technical exchanges for each PRA should also be further deepened in the future.

IV. Adoption of TPM PRA Roadmap (2015–2019) and Discussion of Future TPM

10. The Presidents adopted the TPM PRA Roadmap (2015–2019) after reviewing the revised version, noting that the Roadmap is open to revision during this period.

11. The Presidents acknowledged that the current Lead Coordinating Institutes (LCI) for each PRA will remain as follows: CRAES (Freshwater Pollution, Urban Environment and Eco–city, Chemical Risk and Management); NIES (Biodiversity Conservation, Climate Change, Disaster Environment); NIER (Asian Air Pollution, Dust and Sand

Storm, Solid Waste Management).

12. President SUMI Akimasa pointed out that there is a wide spectrum of environmental issues and a diversity of environmental research interests and needs. He emphasized that each institute has its own mission and is in a unique situation. He also pointed out that the joint research should be done on person–to–person basis. He proposed to discuss a possible reform of the current TPM modality and ways to produce concrete outputs in the PRAs, bearing in mind the principle of TPM, "Friendship, Communication, Cooperation and Win–Win".

13. The Presidents noted that the TPM mechanism is of great significance for promoting regional environmental protection technologies. They emphasized that the three institutes should strengthen the collaboration to achieve "win–win results" in both scientific research and environmental technology progress, based on each country's conditions. They also pointed out that the three institutes should take advantage of each institute's capability in addressing environmental issues of common concern.

V TPM13 International Workshop and Academic Exchange Seminar for the 40th Anniversary of Yunnan Institute of Environmental Sciences: Water Pollution Control Technologies and Ecosystem Health

14. The TPM13 International Workshop and Academic Exchange Seminar for the 40th Anniversary of Yunnan Institute of Environmental Sciences: Water Pollution Control Technologies and Ecosystem Health were held on November 2nd, 2016. It was jointly organized by TPM and the Yunnan Institute of Environmental Sciences (YIES) in order to promote environmental research and the dissemination of scientific findings. The senior leaders and experts from Yunnan Provincial Environmental Protection

Department (YEPD) and YIES attended and gave presentations.

15. Latest scientific research findings were exchanged and discussed in terms of algal blooms and water pollution control in river basins; basin water environmental management and ecological restoration; and water quality criteria and risk assessment.

16. Prof. Sverre VEDAL from Department of Environmental and Occupational Health Sciences, University of Washington, U.S. and Dr. Ganjuur SARANTUYA, Director of Information and Research Institute of Meteorology, Hydrology and Environment, Mongolia were specifically invited to the Workshop to give special technical reports on "Water Pollution and Health".

VI. TPM14

17. President SUMI Akimasa offered to host TPM14 in Southern Tohoku region including Sendai, Japan in 2017.

VII. Conclusion

18. The Presidents expressed their satisfaction with the productive outcome of the TPM13, and emphasized that TPM should be promoted based on the principle of furthering exploration of joint research in PRAs. In this way, scientific exchanges and cooperation can be elevated to the next level, play a more important role in improving environmental quality and promote sustainable development in Northeast Asia.

19. President SUMI Akimasa and President PARK Jin-won expressed their gratitude to CRAES, YEPD and YIES for their active contributions to the success of TPM13 and the warm hospitality extended to all the participants.

中日韩三国环境科学研究院（所）长会议第十四次会议联合公报

2017 年 10 月 27 日　日本筑波

　　应日本国立环境研究所所长渡辺知保的邀请，韩国国立环境科学院院长朴辰远和中国环境科学研究院院长李海生分别率团访问日本筑波，参加了于 2017 年 10 月 24 日至 28 日召开的第十四届中日韩三国环境科学研究院（所）长会议（TPM14）。院（所）长会议于 2017 年 10 月 26 日召开，平行研讨会于 2017 年 10 月 25 日召开，主题为 "国际研讨会——通过评估和管理解决淡水环境问题"。2017 年 10 月 27 日，中韩与会代表参观了日本国立环境研究所总部及其位于霞浦湖的研究站。

　　开幕致辞中，院（所）长们指出三方环境科研合作稳步发展，并希望三方能继续保持友好关系。

　　渡辺知保所长指出，TPM14 会议有许多新面孔加入。他简要介绍了日本国立环境研究所的历史背景、经费预算以及人员情况，着重讲述了东日本大地震后该所的发展情况，包括设立福岛分所和琵琶湖办公室。同时，渡辺知保所长还概述了日本国立环境研究所从事的研究活动。他强调了环境监测、样本储存和供应的重要性，以及环境测量标准化是建立和维护环境研究数据库不可或缺的部分。渡边知保所长还提到，第四个"中长期计划"中启动的新研究活动将促进三方合作。日本国立环境研究所是开展日本本土和国际环境问题研究的核心机构，其研究活动涉及参与环境健康和气候变化相关大型长期项目。韩国国立环境科学院院长朴辰遠强调了 TPM 机制对推动亚洲环境合作的重要性，并分享了韩国国立环境科学院 2016 年取得的成就，以及 2017 年的研究目标：（1）开展环境研究以保护公众健康；（2）帮助人们体会到生活中的切实福利；（3）为新环境政策开展务实研究；（4）响应未来以及全球的环境需求。此外，朴辰遠院长认为三方在 TPM13 上一致同意采纳的《TPM 优先领域合作路线图（2015—2019）》详细地阐述了未来四年的研究方向。他期待路线图将引导产出积极成果，并督促中日韩三国为下一代的福祉继续开展务实合作研究。

中国环境科学研究院院长李海生介绍道，近年来，尤其是近五年来中国的生态环境保护从认识到实践发生了历史性、变革性和全局性变化，生态文明建设取得了显著成效。为改善环境质量，中国以前所未有的力度开展污染控制并且实施中央环保督察制度。李海生院长指出三方应继续合作，保持 TPM 机制的活力。他建议未来要进一步完善联合研究项目、就优先合作领域建立定期研讨会机制，加强并深化三方务实合作，为科学家，特别是青年科学家和研究人员提供更多会面和交流机会。李海生院长进一步呼吁三方尽全力改善东北亚地区环境质量，推动 TPM 机制，将其打造为亚洲乃至全球环境保护科研和创新的标志性平台。

院（所）长们对三方自 TPM13 以来的研究进展和九个优先合作领域的研究活动报告表示认可。作为本届 TPM 平行研讨会主题，会上就"废物管理研究"

方面的最新进展进行了分享。

TPM14 是中国环境科学研究院新任院长和日本国立环境研究所新任所长上任后参加的第一届 TPM 会议。院（所）长们秉持"友谊、交流、合作、共赢"的基本原则，从新视角探讨了 TPM 机制的未来。院（所）长们认为，在科学研究和实际规模方面，当前环境问题变化更大、种类更多。在此情形下，各学科、各研究机构间开展国内和国际合作变得越发必要。院（所）长们认为，作为中日韩三国主要环境研究机构的领导人之间的会议，TPM 能够助力加强三方合作。

基于以上共识，院（所）长们原则上同意开展广泛、实质的 TPM 机制改革，包括寻求 TPM 潜在的新角色，以便在加强现有研究合作的同时，进一步推动三方合作。院（所）长们一致认为通过分享国内国际的合作研究活动经验，TPM 将有助于解决亚洲乃至全球环境问题。院（所）长们对优先合作领域活动产出务实成果的必要性表示关切，并同意进一步讨论弹性机制，开发三方合作研究项目。三方议定改革内容包括：缩短 TPM 会期；重组 TPM 日程及其平行研讨会；由联络员定期安排优先合作领域学术会议；为年轻科学家 / 研究员提供更多机会参加 TPM 活动并展示研究成果。院（所）长们指示工作组筹备改革计划并于下一届 TPM 讨论通过。院（所）长们一致同意，下一届 TPM 应提供《TPM 优先领域合作路线图（2015—2019）》的审查报告。

朴辰遠院长指出了三国在大型出生队列研究领域的最新进展，提议就保护儿童健康，避免儿童暴露于有害化学与环境污染物，设立新的优先合作领域。李海生院长和渡边知保所长对该提议表示关切，并指出需要认真考虑合作的可行性。院（所）长们同意相关领域专家就此提议作进一步讨论。

朴辰遠院长提议于 2018 年在韩国釜山举办 TPM15。

最后，院（所）长们对此次会议所取得的丰硕成果表示赞赏。朴辰遠院长和李海生院长对渡边知保所长及其同事的热情款待和成功主办 TPM14 所作的努力表示衷心的感谢。

The Fourteenth Tripartite Presidents Meeting among NIES, NIER and CRAES Joint Communiqué

October 27, 2017, Tsukuba, Japan

At the invitation of President Watanabe Chiho of the National Institute for Environmental Studies (NIES) of Japan, Presidents Park Jinwon of the National Institute of Environmental Research (NIER) and President Li Haisheng of the Chinese Research Academy of Environmental Sciences (CRAES), heading their delegations, visited Tsukuba, Japan and attended the Fourteenth Tripartite Presidents Meeting (TPM14) from October 24 to 28, 2017. The meeting of the Presidents was held on October 26, 2017 while its parallel workshop entitled "International Workshop – Solving Freshwater Environmental Problems by Assessment and Management" was held on October 25, 2017. The delegates visited NIES Headquarters and its research station at Lake Kasumigaura on October 27, 2017.

In their opening speeches, the Presidents noted the steady development in environmental research collaboration among NIES, NIER and CRAES and expressed their expectation on the continued friendly relationship.

President Watanabe Chiho noted that many participants are new to TPM14. He introduced NIES by outlining the history, budget and personnel with particular reference to the development after Great East Japan Earthquake including the establishment of Fukushima Branch and Lake Biwa Branch Office, as well as the overview of NIES' research activities. He also emphasized the importance of environmental monitoring, storage and provision of samples, and the standardization of environmental measurement which is indispensable to establish and maintain the databases for environmental research. He further mentioned a new research–associated activity launched in the Fourth Medium–and–Long–Term Plan to promote collaboration. NIES serves as a core institute for the domestic and/or international activities which involve long–term and large–scale projects on environmental issues including environmental health and climate change.

President Park Jinwoo of NIER emphasized the importance of TPM in promoting environmental cooperation in all of Asia. He also shared the achievements made by NIER in 2016 and the research goals for this year: conducting environmental research to protect people's health; helping people feel the real benefits in life; carrying out practical research for a new environmental policy; and responding to future and global environmental demands. Furthermore, the president mentioned the agreement made in TPM13 by all three institutions to adopt the "TPM PRA Roadmap" that elaborates on the directions of research that will be taken in the next four years. He expressed his expectations for the positive outcomes of the roadmap and encouraged the three countries to continue with the cooperative and substantial research for the well–being of the next generation.

President Li Haisheng of CRAES introduced that in recent years, especially in the past five years, China's ecological and environmental protection have gone through historical, transitional and overall changes from understanding to practices. Remarkable results have been achieved in ecological civilization construction. China is undertaking unprecedented efforts for pollution control and central government inspection for environmental quality improvement. He pointed out that the three institutes should continue to cooperate for maintenance of the vitality of TPM. He proposed to further strengthen and deepen the pragmatic cooperation among the three institutes by refining joint research projects and establishing a regular seminar mechanism on PRAs, providing more opportunities for the scientists, especially young scientists/researchers to meet and exchange. He further called upon the three institutes to make their utmost efforts to improve the environmental quality in Northeast Asia and promote the TPM, making it a landmark platform for scientific research and innovation in environmental protection in Asia and even in the world.

The Presidents acknowledged the research progress of three institutes since TPM13 and the reports on activities of the nine PRAs. Three institutes' latest developments in the waste management research have been shared at the meeting as the special topic of this TPM.

TPM14 is the first meeting after new presidents came into their positions at CRAES and NIES. The Presidents discussed the future of TPM with new and fresh viewpoints while keeping the principle of "Friendship, Communication, Cooperation, and Win–Win" in mind. They shared the awareness that the environmental problems today are of a greater and broader variety in terms of both scientific spectrum and physical scale. Circumstances increasingly require collaboration among various disciplines and research institutions both domestically and internationally. The Presidents also shared the understanding that the TPM, the forum of the focal points of the three leading environmental research institutes in their own countries, has the capacity to contribute to enhance such collaboration.

Based on this common understanding, the Presidents agreed in principle to make extensive and substantial reform of TPM including the pursuit of TPM's possible new roles to promote the collaboration in addition to enhancing its existing research cooperation. The Presidents shared the idea that TPM could contribute to solve environmental problems in Asia and the world through sharing the experiences from domestic and international collaborative research activities in particular. The Presidents expressed their concerns on the needs to produce practical results of PRA activities and agreed to discuss further on a flexible mechanism for developing a collaborative research project among three institutes. Agreed reforms include shortening the TPM duration; reorganizing TPM program and its parallel workshop; arranging regular academic meetings of PRAs organized by focal points; increasing the opportunity for young scientists/researchers to participate and present their research outcomes in TPM activities among others. The Presidents instructed the Working Group to prepare a reform plan to be discussed and adopted by the next TPM. The Presidents also agreed that a review report of the TPM Roadmap (2015–2019) should be provided at the next TPM.

Noting a recent development of the large–scale birth cohort study in the three countries, President Park proposed to establish a new PRA relating to the protection of children's health from environmental exposure to hazardous chemicals and environmental pollutions. Presidents Li and Watanabe shared their concerns about this proposal, and pointed out the need for a careful consideration of feasibility of collaboration. The Presidents agreed to continue the discussion on this proposal among experts.

President Park offered to host the fifteenth TPM in Busan, Korea in 2018.

Finally, the Presidents expressed their appreciation of the fruitful outcomes of the meeting. President Park and Li extended their sincere gratitude to President Watanabe and his staff members for their hospitality and successful organization of the TPM14.

中日韩三国环境科学研究院（所）长会议第十五次会议联合公报

2018 年 10 月 31 日　韩国釜山

应韩国国立环境科学院张伦硕院长的邀请，中国环境科学研究院院长李海生和日本国立环境研究所所长渡边知保分别率团赴韩国釜山，参加了于 2018 年 10 月 29 日—11 月 2 日举行的第十五届中日韩三国环境科学研究院院（所）长会议（TPM15）。

在会议开幕式上，张伦硕院长向中国环境科学研究院和日本国立环境研究所代表团表示欢迎，并向釜山市政府表示感谢。同时，对工作组成员为组织本次会议所作的努力表示感谢。张伦硕院长指出，合作是解决国内和国际环境问题的关键，并强调 TPM 为三国环境科研院（所）研究员和科学家提供合作平台所发挥的作用。他表示地球同步环境监测光谱仪（GEMS）可在观察东北亚地区空气污染物等方面发挥作用，卫星数据或可共享用于未来合作中。

李海生院长表示，中国政府高度重视生态环境保护工作，中国的环保意识、污染控制力度、环保机构、环保制度和执行情况都有大幅提升和改进。他介绍道，中国重组了生态环境部，并为打赢污染防治攻坚战启动了七大战役。李海生院长强调中国迫切需要提升生态环境保护的科技实力。因此，三国环境科研院（所）应在三方财政支持的情况下，在 TMP 框架下创造更多沟通交流机会。李海生院长建议三国环境科研院（所）全力协作，共同开发"中日韩＋"模式，推进务实合作，不断推进 TPM 成果宣传，为东北亚地区及全球共同繁荣作出贡献。

渡边知保所长表示非常高兴能在 TPM15 上见到熟悉的面孔，他对新成员表示欢迎，同时特别欢迎了韩国国立环境科学院新任院长张伦硕。渡边知保所长简要介绍了日本国立环境研究所的历史背景、经费预算、人员结构、分支机构和设施以及研究项目概况。同时，渡边知保所长宣布继 2009 年温室气体监测卫星 GOSAT 发射后，温室气体监测卫星 GOSAT-2 已于 2018 年 10 月 29 日成功发射。此外，渡边知保所长指出，日本国立环境研究所作为参与日本本土和全球环境问题（包括环境健康和气候变化）大型长期项目的核心机构，将于 2018 年 12 月 1 日成立日本环境与儿童研究以及气候变化适应中心。

2018 年 10 月 30 日，TPM15 主会召开。会上，院（所）长们分享了各自院所的研究活动近况，并就未来应基于共同利益在空气污染、水污染、气候变化和环境健康领域开展研究合作进行了探讨，重点讨论了关于如何推动并加强环境健康研究的具体设想。

院（所）长们讨论了工作组准备的《TPM 优先领域合作路线图（2015—2019）》审查报告，并一致认为尽管其中部分合作进展顺利，但由于时间紧迫和预算限制，还不足以起到加强合作的作用，且未能完全发挥其预期作用。基于以上情况，院（所）长们一致同意 2019 年以后不设优先领域合作路线图。

院（所）长一致认为，随着当前环境和社会情况的变化，应该为 TPM 寻找合适的方法以保持平台可持续发展。院（所）长们一致认可开展人员交流的必要性，特别是培养青年研究人员和／或科学家围绕共同利益开展研究。另外，他们认为也有必要宣传 TPM 研究成果，以支持政策制定。

经过深入讨论和充分考虑，院（所）长们认为 TPM 改革方案将以 2019 年为起始年，从 TPM16 开始实施。TPM 改革方案见附件一。

2018 年 10 月 31 日，釜山广域市保健环境研究院联合主办了主题为"固体废物管理与处置现状和前景"的平行研讨会。院（所）长们对釜山广域市保健环境研究院表示衷心的感谢，并表示他们热切希望且愿意通过积极讨论中日韩三国的问题、经验、技术和措施，进一步加强固体废物管理领域的合作。

李海生院长提议于 2019 年秋在中国杭州举办 TPM16，具体会期将由工作组讨论决定。

TPM15 期间，三国环境科研院（所）参会代表们分别于 2018 年 10 月 29 日和 2018 年 11 月 1 日，至釜山绿色能源有限公司和釜山广域市保健环境研究院以及庆州国家公园和洛东江河口生态中心进行了实地调研。

最后，院（所）长们对此次会议所取得的成果表示赞赏。李海生院长和渡边知保所长对张伦硕院长的热情款待和韩国国立环境科学院的同事们为成功组织 TPM15 作出的贡献表示感谢。

The Fifteenth Tripartite Presidents Meeting among NIER, CRAES and NIES Joint Communiqué

October 31, 2018, Busan, Korea

At the invitation of President CHANG Yoon Seok of the National Institute of Environmental Research (NIER) of Korea, delegations led by President LI Haisheng of the Chinese Research Academy of Environmental Sciences (CRAES) of China and President WATANABE Chiho of the National Institute for Environmental Studies (NIES) of Japan visited Busan, Korea to attend the Fifteenth Tripartite Presidents Meeting (TPM15) from October 29 to November 2, 2018.

In the opening session, President CHANG Yoon Seok welcomed delegates from CRAES and NIES, and expressed his gratitude to Busan City government. He also extended his appreciation to Working Group (WG) members for their hard work in making arrangements for the meeting. He pointed out in his speech that cooperation is the key to addressing environmental problems at domestic and international level. Hence, he made an emphasis on the role of TPM in providing the cooperation platform for researchers and scientists among the three institutes. He mentioned that Geostationary Environment

Monitoring Spectrometer (GEMS) will play a role in observing air pollutants and its relevant in the North East Asia region. The satellite data could be shared for joint cooperation in the future.

President LI Haisheng remarked that the Chinese government attaches great importance to ecology and environment protection, with the environment awareness, pollution control efforts, environmental institutions, regulation and enforcement, as well as achievement in ecological civilization be greatly improved. He introduced China's restructure of the Ministry of Ecology and Environment (MEE) and the launch of Seven Significant Campaigns for win the Battle Against Pollution. He emphasized that there is urgent need for China to strengthen the science and technology for ecology and environment protection. Thus, more chances for communication and exchange under TPM framework should be created with tripartite financial supports. He proposed the three institutes to work together and jointly develop the "China–Japan–South Korea Plus" model, under which pragmatic cooperation should be promoted, with continuous efforts made in the dissemination of TPM fruits, so as to contribute to the common prosperity of Northeast Asia and the world.

President WATANABE Chiho in his Keynote Speech greeted that he was pleased to see familiar faces and welcomed new members at TPM15, especially welcoming the newly appointed President of NIER, Dr. CHANG Yoon Seok. He introduced NIES by outlining the history, budget and breakdown of staff, Branch and facilities, as well as the overview of NIES' research programs. He was also pleased to announce that GOSAT–2, the successor of Greenhouse gases Observation SATellite (GOSAT) launched in 2009, was successfully launched on October 29, 2018. Furthermore, NIES serving as a core institute for both domestic and global activities involving long–term and large–scale projects on environmental issues such as environmental health and climate change, he made particular references to GOSAT-2, Japan Environment and Children's Study (JECS) and Climate Change Adaptation Center at NIES will be established on December 1, 2018.

In the TPM15 Main Meeting held on October 30, 2018, Presidents shared recent developments of research activities in the three institutes and discussed future research collaborations which could be undertaken in the areas of air pollution, water pollution, climate change and environmental health based on common interests. Specific ideas on how to progress and enhance their cooperation on environmental health research were intensively discussed.

The Presidents discussed the Priority Research Area (PRA) Roadmap 2015–2019 Review Report prepared by WG. The Presidents shared that although certain cooperation was successfully progressed under PRA Roadmap 2015–2019, however, it is not enough for enhancing collaboration and did not consummately accomplish its expected roles due to obstacles of tight schedule and budget limitation. Based on these situations, the Presidents shared the understanding that no further PRA Roadmap after 2019.

The Presidents also shared the understandings that as situations in the environment and society are changing and that TPM should seek sustainable ways to keep the pace of platform development. The Presidents appreciated the ideas about the necessity of personnel exchange especially for the cultivation of young researchers and/or scientists to conduct research with common interests. Another necessity is the advocacy of TPM findings to support policy making.

The Presidents also acknowledged that TPM Reform Plan will be implemented from TPM16 in the year of 2019 as the starting year, after thorough discussion and consideration. The TPM Reform Plan is attached in Annex I.

The parallel workshop on "Current Status and Future of Solid Waste Management" was co-hosted by BIHE on October 31, 2018. The Presidents expressed their sincere gratitude to BIHE, and expressed their passion and willingness to further enhance on the cooperation of waste management through active discussions on the three countries' issues, experiences, know-how and measures.

President LI offered to host TPM16 in Hangzhou, China in 2019 autumn. The specific dates for TPM16 will be discussed and confirmed by WG.

During TPM15, the delegates visited the BIHE and Busan Green Energy Co. on October 29, 2018, as well as Gyeongju National Park and Nakdong Estuary Eco Center on November 1, 2018.

Finally, the Presidents expressed their satisfaction with the outcomes of the meeting. President LI and President WATANABE extended their sincere gratitude to President CHANG for his hospitality and also their appreciation to all of NIER members who contributed to the successful organization of TPM15.

中日韩三国环境科学研究院（所）长会议第十六次会议联合公报

2019 年 10 月 30 日　中国杭州

　　应中国环境科学研究院（CRAES）院长李海生的邀请，日本国立环境研究所（NIES）所长渡边知保和韩国国立环境科学院（NIER）院长张伦硕率团参加了于 2019 年 10 月 28 日至 11 月 1 日在中国杭州举办的第十六届中日韩三国环境科学研究院院（所）长会议（TPM16）。浙江省生态环境科学设计研究院院长杨斌作为当地代表及观察员出席了此次会议。在 2019 年 10 月 30 日 TPM16 主会召开前，"东北亚地区大气污染物长距离传输"平行研讨会（PWS）于 2019 年 10 月 29 日召开。2019 年 10 月 31 日日韩与会代表参观了安吉余村和塘溪镇。

　　李海生院长在 TPM16 主会的开幕式上强调，中国生态文明建设实现了历史性进展，并指出，自 2018 年以来，污染防治攻坚战全面打响，中国生态环境质量得到了明显改善。李海生院长还介绍了中国在大气重污染成因与治理攻关项目和长江生态环境保护修复联合研究工作中取得的进展及显著成效。为了有效解决大气污染和长江保护的关键问题，CRAES 作为牵头机构在中国采用了"1+X"和"边研究、边产出、边应用、边完善"的科研模式。同时，CRAES 还继续开展东北亚区域大气污染物跨境传输研究。李海生院长对 CRAES、NIES 和 NIER 三方的环境合作表示高度赞赏。TPM16 作为 TPM 机制改革实施以来的第一次会议，他建议三国环境科研院（所）应进一步凝聚共识，深化合作，探索合作机制，通过"中日韩＋"模式下科研项目和能力建设的统筹规划，促进 TPM 会议为改善东北亚区域环境质量贡献科技力量。

渡边知保所长对李海生院长及其工作人员邀请 NIES 代表团到中国杭州这一历史名城参加 TPM16 表示衷心的感谢。他指出，三国环境科研院（所）都认识到社会对空气、水污染控制及日益重要的环境健康问题的需求。NIES 也尝试与研究机构或利益相关者，如其他公共组织、学术机构、私营公司、非政府组织和社会，进行双向交流来满足各种需求，从他们的观点中学习和获取经验，以进一步加强研究工作。通过由浅入深的研究，NIES 致力于解决社会所关心的新问题，如气候变化。NIES 已于 2018 年 12 月根据《气候变化适应法案》成立了气候变化适应中心（CCCA），并积极与地方政府和其他国内外组织展开合作。正如各国的人口转型始于不同时期，但最终会展现出相似的人口结构。同样地，环境问题出现在不同的时期，但最终会集中在共同的问题上：环境污染和全球环境问题。此外，渡边知保所长还强调了三国环境科研院（所）合作的重要性。

张伦硕院长对于三国环境科研院（所）能在百忙之中在中国会面表示高兴与感激。他表示，研究人员在平行研讨会上的深入讨论给他留下了积极的印象，并确信研究人员的热情将助力三国环境科研院（所）加强合作。他表示，TPM16 作为 TPM 改革计划的开局之年取得了成功 TPM16 为改革后的 TPM 机制起了一个很好的开端。相应地，他希望 TPM 会议时间的缩减能够成为未来鼓励研究人员积极参与的良好契机。张伦硕院长在其主旨演讲中强调了 NIER 为公众提供准确、易懂的环境信息的作用。为了发挥这一作用，他安排了四个特别工作小组，分别集中研究细粉尘（颗粒物）、淡水中的藻华、微塑料和新兴污染物。他认为，由跨学科研究人员组成的特别工作小组将为 CRAES、NIES 和 NIER 的研究人员提供科学灵感。最后，张伦硕院长确信，已持续了 16 年之久的三国环境科研院（所）合作关系将会继续保持下去。

三国环境科研院（所）长对 TPM16 务实合作导向的 TPM 改革进展及 TPM15 以来三国环境科研院（所）的研究进展表示肯定。作为 TPM 改革规划的一部分，三国环境科研院（所）提交了关于 TPM16 平行研讨会的报告，以及空气、水、气候变化和环境健康这四个潜在研究领域（PRAs）的项目提案。三国环境科研院（所）长认为，推进 TPM 改革应以 TPM 框架下的具体合作成果为重点，循序渐进地进行。院（所）长们在充分考虑专家意见和各院（所）的具体情况后，就 PRAs 的项目提案进行了讨论，最终同意在加强气候变化交流的同时，围绕以下三个方面开展联合科学研究：

a. 区域大气污染治理，包括提高空气质量模型的预测精度；

b. 水生态保护经验与技术交流；

c. 环境健康。

三国环境科研院（所）长表示，他们期望看到 CRAES、NIES 和 NIER 在联合研究方面开展更为具体的合作，并进一步加强三国环境科研院（所）之间的沟通。

三国环境科研院（所）长一致认为，随着技术和人员交流的需求日益增加，青年科学家和研究人员需要更加积极地参与到 TPM 会议中，以此扩大年轻人的视野。院（所）长们同意共同发展"中日韩 +"的合作模式。三国环境科研院（所）长一致认为，加强机构间科学活动管理系统的信息和经验交流这一举措是可行的。

TPM16 中举行的平行研讨会议主题是三国环境科研院（所）所共同关注的。在平行研讨会议上，与会者就东北亚地区大气污染物长距离输送问题研究进行了深入讨论。三国环境科研院（所）长强调了在大气污染防治方面开展合作的重要性，以改善东北亚区域环境质量。

渡边知保所长提议 2020 年在日本筑波（Tsukuba）和仙台（Sendai）举办 TPM17 会议，具体日期将由工作层进行讨论和确认。

最后，三国环境科研院（所）长对 TPM 改革实施后的第一年会议所取得的丰硕成果表示赞赏，同时向浙江省生态环境厅和浙江省生态环境科学设计研究院对于 TPM16 的支持和贡献表示感谢。渡边知保所长和张伦硕院长对李海生院长及其工作人员的盛情款待和 TPM16 会议的成功召开表示诚挚的谢意。

The Sixteenth Tripartite Presidents Meeting among CRAES, NIES and NIER Joint Communiqué

October 30, 2019, Hangzhou, China

At the invitation of President LI Haisheng of the Chinese Research Academy of Environmental Sciences (CRAES) of China, President WATANABE Chiho of the National Institute for Environmental Studies (NIES) of Japan and President CHANG Yoon Seok of the National Institute of Environmental Research (NIER) of Korea, heading their delegations, visited Hangzhou, China and attended the Sixteenth Tripartite Presidents Meeting (TPM16) from October 28 to November 1, 2019. As a local representative, Mr. Yang Bin, President of the Environmental Science Research & Design Institute of Zhejiang Province attended TPM16 as an observer. The parallel workshop (PWS) on "Longrange Transboundary Air Pollutants in Northeast Asia" was held on October 29, 2019 prior to the TPM16 Main Meeting on October 30, 2019. The delegates also visited Yucun Village of Anji County and Tangxi Town on October 31, 2019.

In the opening session of the TPM16 Main Meeting, President LI Haisheng emphasized the historic progress of China's ecological civilization. He pointed out that China's environment quality has been greatly improved as the "Battle against Pollution" advanced since 2018. He introduced the smooth progress and remarkable results of China's key projects on the causes of heavy air pollution and key control technologies, and the joint research on the ecology environmental protection of the Yangtze River. As the leading agency, CRAES adopted the "1+X" model with "research and output, application and improvement" platform in China to effectively tackle the key problems of air pollution and Yangtze River conservation. The efforts by CRAES studying the Long–range Transboundary Air Pollutants in Northeast Asia (LTP) was also stretched out. President LI Haisheng highly appreciated the tripartite environmental collaboration among CRAES, NIES and NIER. As the first year of TPM reform's implementation, he suggested that the three institutes should further build consensus, deepen the cooperation and explore cooperation mechanism, by integrated planning of both scientific research projects and capacity building under the "China–Japan–Korea Plus " model so as to promote the TPM's contribution to the ecology and environment in Northeast Asia.

President WATANABE Chiho expressed his sincere gratitude to President LI and all of his staff for inviting NIES delegation over to such a splendid historic city to hold TPM16 in Hangzhou, China. He noted that all three institutes acknowledged the society's demands and needs on air and water pollution control, as well as environmental health issues which are becoming increasingly important. NIES' research also tries to meet various demands and needs through bidirectional communications not only within the research communities but also with stakeholders including other public organizations, academic institutions, private companies, NGOs and the society to learn and gain new aspects from their viewpoints to enhance further studies. Through basic to advanced research studies, NIES is also working on the society's demands and concerns on emerging issues which needs to be tackled fairly quickly such as climate change. NIES established Center for Climate Change Adaptation (CCCA) in December 2018, based on Climate Change Adaptation Act, to work actively with local governments and other domestic/overseas actors. Demographic transition started in various period of time but converged on similar demographic structure across the countries. Similarly, environmental issues occur in different period of time but will be converged on common issues along the way: environmental pollutions and global environment issues. Moreover, he emphasized on the importance of the cooperation among the three institutes.

President CHANG Yoon Seok expressed his gratitude and pleasure to meet CRAES and NIES with busy schedule in China. He had a positive impression from the deep discussion by researchers in Parallel Workshop. He made sure that their passion would be stepping stones towards better cooperation among the three institutes. He expressed that TPM16 was successful as the starting year based on TPM reform plan. Accordingly, he hoped that the shortened duration of TPM in the future can be an opportunity to encourage the participation of researchers. President CHANG in his keynote speech emphasized the role of NIER providing accurate information properly, which is easy to understand, on environmental issues for the public. As part of efforts playing the role, he arranged four task force teams intensively focused on fine dust (particulate matter), algal bloom in freshwater, micro–plastic and emerging contaminants respectively. He believed that the task force teams composed of interdisciplinary researchers would give scientific inspiration to the researchers of CRAES, NIES and NIER. Last but not least, President CHANG was sure that great partnership among the three institutes will be maintained as the three institute has lasted the cooperative relation for 16 years.

The Presidents acknowledged the pragmatic cooperation oriented TPM reform progress of TPM16 and the research progress of the three institutes since TPM15. As part of the TPM reform, the report on TPM16 PWS as well as project proposal of four Potential Research Areas (PRAs), including air, water, climate change and environmental health, were presented. The Presidents understood that the TPM reform should be carried out step by step, focusing on the tangible cooperation results under TPM framework. After discussion on the PRAs project proposals with full consideration of experts' comments and each institute's particular situation, the Presidents agreed that, while enhancing the communication on climate change, joint scientific research would be conducted focusing on the following three aspects:

a. regional air pollution control and treatment, including the improvement of prediction accuracy of air quality model;

b. experience and technology exchange of water ecological protection; and

c. environmental health.

The Presidents expressed their expectation to see more concrete cooperation on the joint research and further communication among CRAES, NIES and NIER.

The Presidents shared the idea that there are growing demands of technical and personal exchange, recalling for more involvement of young scientists and researchers, so as to expand the vision of young people. They agreed to jointly develop the "China–Japan–Korea Plus" model of cooperation.

The Presidents shared the view that enhanced information and experience exchange in the management system for institutional scientific activities would be feasible.

The theme of PWS for TPM16 was based on the common concern among the three institutes. In the PWS, the participants had deepened discussion on longrange transboundary air pollutants in Northeast Asia. The Presidents emphasized the importance to work together on the air pollution control for the improvement of the environmental quality of the Northeast Asia.

President WATANABE Chiho offered to host TPM17 in Tsukuba and Sendai, Japan in 2020. The specific dates for TPM17 will be discussed and confirmed by the working group.

Finally, the Presidents expressed their appreciation of the fruitful outcomes of the meeting as the first year implementing TPM Reform Plan. The Presidents also extended their gratitude to the Zhejiang Ecology and Environment Bureau and the Environmental Science Research & Design Institute of Zhejiang Province for their supports and contributions to TPM16. President WATANABE Chiho and President CHANG Yoon Seok extended their sincere gratitude to President LI Haisheng and his staff members for their hospitality and successful organization of the TPM16.

中日韩三国环境科学研究院（所）长会议第十七次会议
联合公报

2020 年 12 月 16 日

应日本国立环境研究所（NIES）所长渡边知保的邀请，韩国国立环境科学院（NIER）院长张伦硕和中国环境科学研究院（CRAES）院长李海生率团参加了于 2020 年 12 月 16 日举办的第十七届中日韩三国环境科学研究院院（所）长会议（TPM17）。三国环境科研院长参加的 TPM17 是 TPM 历史上首次线上会议。

渡边知保所长在 TPM17 主会的开幕致辞中指出，受新冠疫情影响，此次 TPM17 的举办方式进行了相应转变。他分享了 NIES 新冠疫情下的应对措施。根据日本政府相关要求，NIES 取消、暂停并缩减了研究活动。日本政府也要求人们应佩戴口罩。渡边知保所长对李海生院长在 2020 年 3 月日本最急缺口罩时，及时寄口罩给 NIES 深表谢意。随着新冠疫情持续在全球范围传播，NIES 致力于更多地使用网络通信设备，并鼓励员工居家办公。通过此次实践证明，即使有大量员工居家办公，NIES 仍可以作为研究机构正常运行。渡边知保所长还分享了新冠疫情对气候变化影响的最新研究发现。NIES 研究人员发现，由于新冠疫情影响，2020 年日本 CO_2 排放量显著减少，但大气中的 CO_2 浓度仅呈现微弱的下降。这些发现促使 NIES 研究人员对绿色复苏进行了讨论。随着 NIES 的第五个"五年计划"于 2021 年 4 月生效，NIES 将致力于加强全球研究与本地研究之间的关联度，夯实基础研究和巩固环境监测技术，并促进与社会的对话与合作。渡边知保所长在开幕致辞的最后表示，希望会议当天三方可以畅所欲言、积极讨论。

在新冠疫情日益严重的情况下，张伦硕院长对于仍可通过线上的方式与 NIES 和 CRAES 代表团会面表示高兴与感激。面临当前的严峻时期，张伦硕院长表示对 NIER 近期所取得的成果十分骄傲，并进行了相应介绍。第一个里程碑是 NIER 于 2020 年 2 月发射的环境卫星：地球同步环境监测光谱仪（GEMS）。这颗卫星首次实现了对韩国和整个亚太地区的监测，意味着韩国将与欧洲、美国共同持续监测大气环境。为致力于守护蓝天，空气质量监测站已与当地研究机构联网，分区域对空气污染现象和成因进行分析。第二个里程碑是环境健康领域，韩国已开展了韩国国家环境健康调查，旨在调查易受污染地区对人们尤其是儿童健康的影响。此外，NIER 还开展了环境有害因素风险评估相关研究，并在此研究基础上提出了科学措施。第三个里程碑是对废弃物问题采取的行动。为了安全有效地处理日益增加的废弃物，NIER 开展了针对废塑料回收的全过程研究，并同时就进入焚化设施的不可燃材料相关有效管理措施进行了研究，以减少焚化量。在开幕致辞的最后，张伦硕院长总结道，此次会议为大家提供了一次回顾 TPM 历史的好机会，是一次有意义的线上会议。展望未来，他希望三国环境科研院（所）可以在新平台上进行更加频繁的交流。

李海生院长就渡边知保所长和 NIES 员工在这一特殊的年份精心组织召开了 TPM17 会议表示衷心的感谢。他介绍了中国统筹推进新冠疫情防控和经济社会发展复苏所作出的齐心努力、"十三五"规划时期的中国生态环境保护进展及其重点任务。他强调，"十三五"规划纲要确定的生态环境领域 9 项约束性指标，有 8 项已提前完成，各项目标任务预计将于 2020 年年底全面完成。李海生院长进一步介绍了 CRAES 为打赢打好污染防治攻坚战所提供的科技支撑，包括多举措助力打赢新冠疫情防控阻击战；增强疫情防控的环境保护应急能力；大气重污染成因与治理攻关项目的显著成效；以及长江生态环境保护修复联合研究的进展。李海生院长建议三方机构可以将"碳中和"作为三方合作的切入点，通过探索进一步务实合作、多渠道开展合作，更好地改善东北亚生态环境质量，共建美丽地球。

院（所）长们对三国环境科研院（所）自 TPM16 以来的研究进展、潜在合作领域（PRAs）中气候变化和环境健康领域的书面报告表示肯定。院（所）长们还听取了中日韩三方科技创新体制机制对比研究（以下简称对比研究）的初步成果汇报。

在院长讨论环节，三国环境科研院（所）长分享了他们在多边合作研究方面的经验。他们认可科研人员交流和信息共享对成功合作的重要意义。院（所）长们还就各自机构如何适应国情实现水管理的政策目标进行了交流。

围绕应对气候变化这一全球性问题，三国环境科研院（所）长交流分享了各自进行的有关行动，如 NIES 的应对气候变化中心（CCCA）所开展的相关研究与活动。三国环境科研院（所）长一致认可人工智能（AI）技术在环境科学中的重要性。基于以上认知，院（所）长们建议，针对诸如气候变化适应或人工智能在环境领域的应用等主题之一举行线上研讨论坛。

三国环境科研院（所）长肯定了对比研究的结果，并就尽可能最大努力地促进人员交流和筹集专项资金进行了讨论。他们还建议定期举办线上研讨论坛，以便找到三国环境科研院（所）之间可能的合作主题。三国环境科研院（所）长还同意在接下来的一年中，进行进一步的深入对比研究，并由三方工作层在 TPM18 会议中作汇报。

院（所）长们讨论的另一个议题是 TPM 与其他平台之间的关系，如中日韩环境部长会议（TEMM）。三国环境科研院（所）长共同确认，TPM 会议是三国环境科研院（所）探讨、促进科研合作的独立平台，并也可为各自国家的环境政策作出贡献。

TPM18 会议将在三国环境科研院（所）长的同意下于 2021 年举办。在兼顾新冠疫情的情况下，TPM 工作层将通过电子邮件和在线工具作进一步的具体讨论。

最后，三国环境科研院（所）长对 TPM 第一次线上会议 TPM17 的成功举办表示赞赏，并对会议的成果深表满意。张伦硕院长和李海生院长对渡边知保所长及其工作人员成功组织召开 TPM17 会议表示诚挚的谢意。

The Seventeenth Tripartite Presidents Meeting among NIES, NIER, and CRAES Joint Communiqué

December 16, 2020

At the invitation of President WATANABE Chiho of the National Institute for Environmental Studies (NIES) of Japan, President CHANG Yoon Seok of the National Institute of Environmental Research (NIER) of Korea and President LI Haisheng of the Chinese Research Academy of Environmental Sciences (CRAES) of China, heading their delegations, attended the Seventeenth Tripartite Presidents Meeting (TPM17) on December 16, 2020. The Presidents joined TPM17 online for the first time in the history of TPM.

In the opening speech of the TPM17 Main Meeting, President WATANABE mentioned that COVID–19 affected how to deliver TPM17 and shared NIES' responses to the pandemic. Following instructions from the Japanese Government, NIES canceled, suspended and scaled down research activities. The Government also requested people to wear masks and President WATANABE expressed his deep appreciation to President LI for the masks he sent to NIES in March 2020 when masks were in short supply.

With COVID-19 continuing to spread worldwide, NIES focused on making a greater use of ICT equipment and promoted working from home. The promotion proved that NIES can operate as a research institute even with many staff working from home. President WATANABE also shared recent research findings of COVID-19 impacts on climate change. NIES researchers found that CO_2 emissions decreased markedly in 2020 as a result of the pandemic but atmospheric CO_2 concentrations only showed a faint dive. Such findings have prompted discussion among NIES researchers on green recovery. With NIES' Fifth Five-Year-Plan coming into effect in April 2021, NIES will be focusing on strengthening linkages between global research and local research, consolidating basic research and environmental monitoring, and promoting dialogues and cooperation with the society. President WATANABE closed his opening speech by expressing expectations that active discussions be made today.

President CHANG Yoon Seok expressed his gratitude and pleasure to meet delegation from NIES and CRAES through the online platform amid growing concerns over the COVID-19 pandemic. Under harsh conditions, President CHANG was proud to introduce achievements NIER has made. A first milestone is NIER's launch of environmental satellite, Geostationary Environment Monitoring Spectrometer (GEMS) in February 2020. This satellite can monitor over South Korea and the greater Asia-Pacific region for the first time. Korea stands shoulder to shoulder with the Europe and US to keep monitoring atmosphere environments. In an effort for a blue sky, air quality monitoring stations in operation have networked with local research institutes to analyze the phenomenon and cause of air pollution by region. A second milestone has been made in Environmental Health. The Korean National Environmental Health Survey has been conducted to investigate the effect of areas vulnerable to pollution on people's health especially children. In addition, NIER has conducted research on the risk assessment of environmentally harmful factors, and based on this research has come up with scientific measures. A third milestone is action on the problem of wastes. As part of efforts to deal with increasing wastes in a safe and efficient manner, NIER has conducted research on the entire process of wasted plastics recycling. NIER has also studied efficient management measures for non-combustible materials being carried into incineration facilities in order to decrease the amount of incineration. In the end of keynote speech, President CHANG concluded that this is a good opportunity to look back on the history of TPM and a meaningful online meeting. Going forward, he hopes that three institutes will have frequent meetings on the new platform.

President LI Haisheng expressed his sincere gratitude to President WATANABE for NIES' well organization of TPM17 in such a special year. He introduced that concerted efforts were made in China to advance epidemic prevention and control while recovering the economic and social development. President LI shared the progress achieved in China's ecology and environment protection, including key tasks, during the 13th Five-Year Plan period. He highlighted that the eight out of totally nine obligatory targets for ecology and environment protection set out in the 13th Five-Year Plan were met ahead of schedule, and all targets would be fully completed by the end of 2020. President LI further introduced CRAES' efforts providing scientific and technological support for winning the "Battle against Pollution", including taking multiple measures to support winning the battle against COVID-19 and responding to the urgent needs for environmental protection therein, remarkable results of the Project on the Causes and Control Measures of Heavy Air Pollution, and the progress in the joint research on the restoration of the ecology and environment of the Yangtze River. President LI suggested that the three institutes could take the "carbon neutrality" as an opportunity making breakthrough in tripartite cooperation, with further practical cooperation be explored and multiple channels be developed for better improvement of the ecology and environmental quality in Northeast Asia, so as for our beautiful earth.

The Presidents acknowledged the research progress of three institutes since TPM16 and the reports on activities of Potential Research Areas (PRAs) on Climate Change and Environmental Health. The Presidents also listened to the preliminary results of "Comparative Study of Science and Technology Innovation System and Mechanism of CRAES, NIES, and NIER" (hereinafter referred to as "Comparative Study").

In the Discussion, the Presidents shared their experiences on multilateral research collaboration. They recognized the importance of researchers exchange and information sharing for successful collaborations. The Presidents exchanged the situation of each country on how each institute contributed to achieve the policy target of water management.

Since climate change adaptation is a global issue, the Presidents shared activities on climate change adaptation such as research and activities by the Center for Climate Change Adaptation at NIES. The Presidents shared the same view on the importance of the Artificial Intelligence (AI) technology in environment sciences. Based on these understandings, the Presidents suggested to hold online research forums on one of the themes such as climate change adaptation or the application of AI in the field of environment.

The Presidents acknowledged the findings of the Comparative Study and discussed to make the best possible effort to elaborate personnel exchange and fund raising. They also suggested to hold periodical online research forums in order to find a possible topic of research collaboration among the three institutes. It was also agreed that further in-depth Comparative Study would be conducted in the coming year and the working group level from the three institutes would make a presentation in the TPM18.

Another topic discussed was the relationship between TPM and other platforms such as the Tripartite Environment Ministers Meeting (TEMM) among Japan, Korea and China. The Presidents confirmed that TPM is an independent platform among the three institutes for finding and facilitating research collaborations which can also contribute to environmental policies in their country.

TPM18 will be held in 2021 as agreed by the Presidents. With due consideration to the pandemic situation, the TPM Working Group will discuss further detail using emails and online tools.

Finally, the Presidents expressed their appreciation of the first online TPM17 and their deep satisfaction with the outcome of the meeting. President CHANG and President LI extended their sincere gratitude to President WATANABE and his staff for their successful organization of TPM17.

中日韩三国环境科学研究院（所）长会议第十八次会议联合公报

2021 年 11 月 4 日

应韩国国立环境科学院（NIER）院长金东镇的邀请，中国环境科学研究院（CRAES）院长李海生和日本国立环境研究所（NIES）所长木本昌秀率团参加了于 2021 年 11 月 4 日举办的第十八届中日韩三国环境科学研究院院（所）长会议（TPM18）。考虑新冠疫情影响，此次会议延续 TPM17 会议形式，通过线上召开。

金东镇院长在 TPM18 主会的开幕致辞中向李海生院长、木本昌秀所长以及所有参会者表示诚挚的欢迎。同时，金东镇院长对木本昌秀所长就任日本国立环境研究所所长表示祝贺。当前，全球高度关注气候变化，金东镇院长强调应在实现碳中和目标下开展以未来为导向的研究，这其中，中日韩三国环境科研院（所）的紧密合作至关重要。随后，金东镇院长介绍了韩国国立环境科学院的三个关键研究领域进展。第一个里程碑是大气质量研究进展。NIER 于 2020 年 2 月发射的环境卫星已实现对韩国和整个东北亚地区的大气环境监测。自 2021 年 3 月起，空气质量图像已对大众公开，其大气环境数据的精准度一直在稳步提升。第二个里程碑是环境健康领域，对此，NIER 一直聚焦于保护人类健康。在公共卫生方面，

NIER 持续开展两项主要研究，分别针对易受污染级别和区域。第三个里程碑是废弃物管理。NIER 在此领域开展了大范围的研究，在减少包括塑料在内的废弃物方面付诸极大努力，并同时致力于提高废弃物的回收率。此外，NIER 还建立了"废弃物转能源"和"废弃物资源回收"的基地，朝着协同碳中和的资源循环型社会而努力。

李海生院长对于金东镇院长精心组织召开了 TPM18 会议表示衷心的感谢。他诚挚地祝贺金东镇院长和木本昌秀所长就任新职。李海生院长介绍了中国在"十四五"时期（2021—2025 年）面临的新形势和新挑战。他指出，在过去的五年中，中国的生态环境质量显著提升，人民幸福感大大增加。"双碳"时代与"深入打好污染防治攻坚战"为中国的生态环境保护科技带来了新机遇、新挑战和更高的要求。因此，中国环境科学研究院致力于增强自身科技创新能力和管理支撑能力。李海生院长简要地介绍了过去一年间中国环境科学研究院在五个方面的工作进展，包括建立中韩联合环境研究实验室，开展 $PM_{2.5}$ 与 O_3 协同控制研究，启动碳达峰与碳中和研究，为黄河流域生态环境保护和高质量发展提供基于科学的解决方案与建议，以及持续推动环境基准与标准建设。关于 TPM 机制，李海生院长指出尽管中日韩三国作为近邻同处东北亚，但因各自的发展情况不同，三国仍需面临不同的环境问题和压力。因此，三国环境科研院（所）可在共同兴趣领域探索合作，包括区域和全球的环境问题，诸如气候变化。

木本昌秀所长对金东镇院长及其代表团主办 TPM18 线上会议表达了衷心的感谢，同时也感谢了李海生院长及其代表团在此次会议中的积极参与。他提到，环境问题在近年来越发重要，并介绍了日本国立环境研究所的气候变化适应中心。该中心开展气候变化影响的定量评估和机制研究、气候变化影响评估方法的复杂性研究，以及基于科学预测的气候变化适应策略制定和应用研究。该中心将持续为地方政府、企业和个人提供技术支持，通过简单易懂的方式向他们提供研究成果和信息，并通过上述信息的应用来推广气候变化适应措施。此外，日本国立环境研究所正在计划建立新的区域合作体系，旨在与区域气候变化适应中心合作开展联合研究，通过收集当地站点的信息提供区域支持，并力求将此体系扩大至国家、亚太地区乃至全球。2021 财政年是日本国立环境研究所第五个五年计划的开局之年，其中，推进与本国和国际科研机构之间的合作是上述计划设定的目标之一。木本昌秀所长表示，中日韩三国环境科学院院（所）长会议是日本国立环境研究所最重要的活动之一。木本昌秀所长希望，将来三国环境科研院（所）可以通过合作、竞争和相互促进，成为环境研究领域的典范。

TPM18 会议是日本国立环境研究所及韩国国立环境科学院两位新院（所）长到任以来共同参加的第一次会议。展望世界环境格局，结合当前实现碳中和趋势，院（所）长们对三国环境科研院（所）近来的研究活动进展表示一致认可。他们表示，通过开展"中日韩三国环科院（所）科技创新体制机制深入对比研究"，三国环境科研院（所）加深了对彼此的了解。其中，NIER、CRAES、NIES 分别牵头针对科研管理和科技产出、科技发展战略、人才发展战略开展了深入对比研究。该深入对比研究是 TPM17 会议上"中日韩三方科技创新体制机制对比研究"的延续，由中国环科院提议，三国环境科研院（所）共同完成。

院（所）长们认为"气候变化适应"作为会议主题学术研讨的议题恰逢其时，因为当今世界正大踏步朝着碳中和目标迈进。他们一致认可潜在合作领域（PRA）微论坛展示了三国环境科研院（所）开展合作研究的未来愿景。基于以上理解，秉持"友谊、交流、合作、共赢"原则，他们针对如何加强中日韩三方环科院（所）

合作，共同实现低碳社会开展了讨论。

院（所）长们认可三国环境科研院（所）的紧密合作对于实现低碳和可持续社会至关重要。谈及合作，他们表示他们希望三国环境科研院（所）能够基于先进科技和信息共享，深入开展大气环境潜在合作领域的联合研究。此外，考虑NIER 的水质系统和 CRAES 的长江、黄河保护项目，院（所）长们表示，三国环境科研院（所）可在水环境潜在合作领域找到合作研究点。

院（所）长们特别提到了开展以未来为导向的研究的重要性，尤其是在气候变化适应、人工智能（AI）和大数据领域。面对做好实现碳中和目标的准备，他们对开展气候变化适应的相关合作表示出浓厚兴趣，并强调了以气候变化适应中心为主，开展多元化信息交流的需求。院（所）长们期待日本国立环境研究所设立的气候变化适应信息平台（A-PLAT）和亚太气候变化适应信息平台（AP-PLAT），能在三国环境科研院（所）气候变化领域的合作中起到关键作用。结合未来研究领域，院(所)长们对人工智能与大数据领域可能的合作表示高度期待，并希望能以人工智能、大数据领域可能的合作多方位促进环境研究。

TPM19 会议将于 2022 年由中国环境科学研究院主办。李海生院长表示，在疫情防控允许的条件下，将在中国主办 TPM19 会议。具体会议举办的日期和形式，将由工作层作进一步的讨论和确认。

最后，三国环境科研院（所）长对新冠疫情时期 TPM18（延续 TPM17 的线上形式）的成功举办表示赞赏，他们对会议的成果深表满意。李海生院长和木本昌秀所长对金东镇院长及工作人员成功组织召开了 TPM18 会议，表示了诚挚的谢意。

The Eighteenth Tripartite Presidents Meeting among NIER, CRAES and NIES Joint Communiqué

November 4, 2021

At the invitation of President KIM Dong Jin of the National Institute of Environmental Research (NIER) of Korea, President LI Haisheng of the Chinese Research Academy of Environmental Sciences (CRAES) of China and President KIMOTO Masahide of the National Institute for Environmental Studies (NIES) of Japan, heading their delegations, attended the Eighteenth Tripartite Presidents Meeting (TPM18) on November 4, 2021. The Presidents joined TPM18 online following TPM17 over growing concerns about the COVID–19 pandemic.

In the opening speech of the TPM18 Main Meeting, President KIM gave a warm welcome to President LI Haisheng, President KIMOTO Masahide, and all the participants. President KIM offered President KIMOTO a special congratulation for being appointed as NIES' President. Amid the world keeping a close eye on Climate Change, he made an emphasis on closely cooperating with each other among NIER, CRAES and NIES under the vision of Carbon Neutrality related with a future–oriented research and then shared recent developments in three key research areas achieved by NIER. The air quality was a first milestone in the research progress. In February, 2020, NIER has launched the Environmental Satellite playing an important role in monitoring the atmospheric environment over the North East Asia as well as Korea. Air quality images have been open to the public since March in 2021 while improving the data accuracy of the atmospheric environments. A second milestone is the Environmental Health area. It is the research area that NIER has focused to protect people's health. In an effort for the public health, we, NIER, continue to conduct two main researches for vulnerable class and areas. A third milestone area is the waste management. We have studied a wide range of researches decreasing wastes including plastics in a serious endeavor while increasing the rate of its recycling. In addition, NIER has established the base to turn waste into energy and energy recovery from waste, moving toward a resource circulation society in line with Carbon Neutrality.

President LI Haisheng expressed his sincere gratitude to President KIM Dong Jin for NIER's well organization of TPM18. He congratulated President KIM and President KIMOTO for taking office as the President of NIER and NIES, respectively. President LI introduced the new situation and challenges of the 14th Five-Year Plan period (2021–2025) of China. He pointed out that during the last five-year period, the ecology and environment quality in China has been improved obviously, with people's happiness being increased remarkably. Noting that the era of "Carbon Peaking and Carbon Neutrality" and "In-depth Battle Against Pollution" has brought China's scientific technologies for the protection of ecology and environment with new opportunities, new challenges and higher requirements, CRAES is devoting to strengthen the capacity of science and technology innovation as well as the management support thereof. President LI briefed the past-year working progress of CRAES in five aspects, including the establishment of China–Korea Joint Environmental Research Laboratory, the research of $PM_{2.5}$ and O_3 coordinated control, research on carbon peaking and carbon neutrality, science–based solution and recommendation for the ecology and environment protection and high–quality development of the Yellow River Basin, as well as the development of environmental criteria and standards. With regard to TPM mechanism, President Li commented that although China, Japan, and Korea are situated closely in the Northeast Asia, the three countries need to face different environmental issues and pressures due to different development situations. In this regard, the three institutes could explore cooperation in the area of common interests, including the regional and global issues, like climate change.

President KIMOTO Masahide expressed his sincere gratitude to President KIM and his delegation for hosting TPM18 online, as well as to President LI and his delegations for their active participation. He pointed out how important environmental issues have become in recent years and introduced NIES' Center for Climate Change Adaptation. The center conducts research on the quantitative evaluation of the effects of climate change and the elucidation of its mechanism; the sophistication of climate change impact assessment methods; and the formulation and practice of adaptation strategies based on scientific predictions. The center will continue to technically support local governments, businesses and individuals by providing research outcomes and information in an easy–to–understand manner for them to promote adaptation measures using this information. Furthermore, NIES is planning on to create a system to cooperate with regional climate change adaptation centers to conduct joint research, give regional supports by gathering information at local sites, and to develop them over wider areas–domestically and in the Asia–Pacific region and abroad. Since FY2021 marks the first year of NIES' Fifth Five–Year Plan and that promoting collaborations with domestic and international institutes is one of the goals set, President KIMOTO acknowledged that the Tripartite Presidents Meeting among NIES, NIER, and CRAES is one of the most important activities at NIES. President KIMOTO hopes that the three institutes would serve as a model in the field of environmental research by cooperation, competition and enhancing with each other in the years to come.

TPM18 is the first meeting after new Presidents came into their positions at NIER and NIES. The Presidents recognized the recent developments of research activities after their speech about the world's outlook for the environment, carbon neutrality. They had a better awareness of the three institutes through *In-depth Analysis on the Comparative Study among the Three Institutes*, in terms of scientific research management and science and technology output by NIER, development strategies by CRAES, as well as talent development strategies by NIES. The in–depth analysis herein is a continuation of the study initiated by CRAES and jointly conducted by the three institutes since TPM17.

The Presidents had similar views that climate change adaptation was an appropriate theme for the research focus session as the world is taking a big step toward carbon neutrality. They also reached the common ground that the mini forum of the Potential Research Areas (PRAs) showed a future vision of three institutes' cooperative research points. Based on these understanding, they discussed how to approach a decarbonization society in close collaboration of NIER, CRAES and NIES while keeping the principle of "Friendship, Communication, Cooperation, and Win–Win" in mind.

The Presidents acknowledged that close cooperation among the three institutes is crucial to aim for decarbonization and a sustainable world. In terms of cooperation, they expressed that the three institutes could deepen the joint research in the Air PRA based on advanced technology and information sharing. In addition, they remarked that the three institutes may be able to find cooperative research points in the Water PRA since there were the water quality system of NIER and project on the Yangtze River and the Yellow River protection by CRAES.

The Presidents noted the importance of a future-oriented research such as climate change adaptation, artificial intelligence (AI) and big data. They showed their keen interests in climate change adaptation ahead of preparation for carbon neutrality, emphasizing the need to exchange various information centered-around the Center for Climate Change Adaptation. They expected the Climate Change Adaptation Information Platform (A-Plat) and the Asia-Pacific Climate Change Adaptation Information Platform (AP-Plat) developed by NIES to play a key role in cooperation on climate change with NIER and CRAES. In line with futuristic research areas, the Presidents remarked the possibility of collaborative research in AI and big data fields and showed a high expectation that they would advance environmental studies in various ways.

TPM19 will be hosted by CRAES in 2022. President LI Haisheng offered to host TPM19 in China in 2022 if the situation of pandemic control permits. The specific date and form for TPM19 will be discussed and confirmed by the working group members.

Finally, the Presidents expressed their appreciation of having TPM18 online following TPM17 over concerns of the pandemic and their deep satisfaction with the outcome of the meeting. President LI and President KIMOTO extended their sincere gratitude to President KIM and his staff for their successful organization of TPM18.

中日韩三国环境科学研究院（所）长会议第十九次会议联合公报

2022 年 11 月 24 日

应中国环境科学研究院（CRAES）李海生院长的邀请，日本国立环境研究所（NIES）木本昌秀所长和韩国国立环境科学院（NIER）金东镇院长率团于 2022 年 11 月 24 日参加了中日韩三国环境科学研究院（所）长会议第十九次会议（TPM19）视频会。

在 TPM19 主会开幕式上，李海生院长对木本昌秀所长、金东镇院长以及来自 NIES 和 NIER 的所有与会者表示热烈欢迎，对三国环境科研院（所）的工作层成员克服新冠疫情困难成功举办本次会议所作出的贡献表示感谢。

李海生院长回顾了三国环境科研院（所）的合作，高度评价三方在 TPM 机制下科研工作的战略意义及其在助力东北亚环境治理方面的成就。李海生院长从绿色发展和生态环境保护等方面简要介绍了中国生态文明建设取得的历史性进展。他介绍中国政府始终高度重视生态文明建设，围绕生态环境保护和绿色低碳发展工作开展了一系列根本性、开创性、长远性工作。李海生院长从服务生态环境保护工作顶层设计、助推经济社会发展全面绿色转型和支撑深入打好污染防治攻坚战三个方面总结了 CRAES 在科技创新和管理决策支撑方面取得的成绩。李海生院长提议，在三方共同关心的问题上从三个方面进一步加强合作，一是结合减污降碳协同增效，促进气候变化领域的科研交流和应对措施共享；二是围绕区域大气污染防治、水生态保护和环境健康机理等领域保持广泛合作；三是充分发挥科技创新支撑作用，助力解决实际问题，并加强青年科研人员交流。

木本昌秀所长对李海生院长及其代表团以视频会议形式组织举办 TPM19 表示感谢，对金东镇院长及其代表团的积极参加表示感谢。木本昌秀所长谈到今年是中日邦交正常化 50 周年。这些年来，NIES 与 CRAES 开展了广泛的研究项目合作，为科研人员和政府官员通过国际合作共同努力解决各种环境问题树立了良好的榜样。NIER 也一直是 NIES 的联合研究伙伴，特别是由 NIER 领导的三国间的东北亚大气污染物长距离跨境传输研究，以及环境对人体健康影响的评估研究。并且，他介绍了 NIES 第五个五年计划的目标是促进社会实施，包括对政策制定的贡献。他还谈及 NIES 派员参加联合国气候变化框架公约第 27 次缔约方会议时，在提高世界各地高级官员对降碳和气候适应问题的认识并吸引他们的广泛兴趣方面所作的努力。木本昌秀所长表达了他认为环境问题无国界的观点，同时也认可 TPM 的重要性以及三国环境科研院（所）作为东北亚地区领先机构的作用。最后，木本昌秀所长希望在今天的演讲和讨论中听到更多关于 TPM 如何为未来作出贡献的信息。

金东镇院长强调，国际合作是解决世界各地环境问题的关键。在中日邦交正常化 50 周年，中韩建交 30 周年之际，金东镇院长提到，大气领域的全球研究合作愈加广泛，他对 NIER 和 CRAES 联合研究小组在减少大气污染物和实现碳中和方面取得的丰硕成果引以为傲。金东镇院长表示，在此背景下，NIER、CRAES

和 NIES 应加强其合作伙伴关系，这是解决问题的关键，他并分享了 NIER 的三个战略重点领域。一是扩大面向未来的气候变化研究，通过建立国家气候危机适应中心（NACCC），在国家层面上监测温室气体的排放，迈向无碳社会。二是投入精力发展第四次工业技术融合研究，为人民的生活创造美好环境。最后，金东镇院长强调了环境安全网对公众健康的重要性，他并补充道，NIER 已努力提出了缓解噪声和光污染对人民健康影响的措施。

院（所）长们回顾了中日韩三方的双多边合作历史，以纪念中日邦交正常化50 周年，中韩建交 30 周年。各方一致认为，TPM 机制的建立正是基于良好双边合作所奠定的深厚基础。自 2004 年成立以来，TPM 为促进东北亚地区的可持续发展发挥了积极有效的作用。

院（所）长们听取了 CRAES 从气候变化观测、NIES 从减缓和公众意识以及 NIER 从气候变化的影响和适应等方面所作的"气候变化对比研究"的成果报告。院（所）长们一致认为，三国环境科研院（所）目前的研究成果可用于气候变化领域的深入合作研究，特别是交流和讨论如何实现减污降碳协同增效。院（所）长们指出，"对比研究"是探讨三方共同关注的问题和寻求 TPM 机制下进一步务实合作的有效途径。

院（所）长们批准了潜在合作领域牵头机构的变更机制，即每三年更换一次牵头单位，以激发潜在合作领域的新活力。在新的任期内，大气污染领域牵头单位将从 NIER 更换为 CRAES，环境健康领域牵头单位将从 NIER 更换为 NIES，水环境领域牵头单位将从 CRAES 更换为 NIER，气候变化领域牵头单位仍为 NIES。

院（所）长们听取了新一轮潜在合作领域牵头单位汇报的合作规划，并讨论了未来合作。院（所）长们一致认为，统筹减污降碳协同增效，坚持 $PM_{2.5}$ 和 O_3 协同治理可作为未来大气污染领域的合作重点。在水领域，三方可就先进技术和做法进行更多交流分享。在气候变化领域，三方的气候变化研究中心可以在碳监测和气候变化评估建模以及适应方面加强交流与合作。在环境健康领域，三方有广阔的合作前景，可进一步拓展。最后，院（所）长们建议对 TPM 的进展和成就进行梳理回顾，并在更大范围内公布或向公众公开。

为讨论东北亚地区共同关心的问题，TPM19 的平行研讨会（PWS）的主题是"基于新兴技术的大气观测与源解析"。李海生院长在平行研讨会上作主旨演讲，介绍了中国在习近平生态文明思想指引下开展大气污染防治的经验及成效。木本昌秀所长和金东镇院长对中国在大气污染控制方面取得的成就表示肯定。院（所）长们一致强调，三方可在该领域进一步加强务实合作，因为大气污染并非一个国家的问题，而是一个区域的问题。

木本昌秀所长表示愿意在日本举办 TPM20，具体会期和形式将由工作层讨论和确认。院（所）长们授权工作组在明年进行详细讨论。

最后，院（所）长们对此次会议所取得的成果深表满意。木本昌秀所长和金东镇院长对 CRAES 代表团全体代表为 TPM19 的成功召开所作出的贡献表示感谢。

The Nineteenth Tripartite Presidents Meeting among CRAES, NIES, and NIER Joint Communiqué

November 24, 2022

At the invitation of President LI Haisheng of the Chinese Research Academy of Environmental Sciences (CRAES) of China, President KIMOTO Masahide of the National Institute for Environmental Studies (NIES) of Japan, and President KIM Dong Jin of the National Institute of Environmental Research (NIER) of Korea, heading their delegations, attended the Nineteenth Tripartite Presidents Meeting (TPM19) virtually on November 24, 2022 hosted by CRAES.

In the opening session of the TPM19 Main Meeting, President LI Haisheng extended a warm welcome to President KIMOTO, President KIM, and all the participants from NIES and NIER. He expressed appreciation for the contribution of Working Groups from three institutes to the success of this meeting by overcoming the difficulties caused by the COVID–19 pandemic.

President LI Haisheng spoke highly of the strategic importance of scientific research and its achievements under TPM in supporting the environmental governance in Northeast Asia, by looking back at the tripartite cooperation among three institutes. He briefly introduced China's historic progress towards an ecological civilization in terms of green development and ecological environmental protection. He remarked that the Chinese government has always attached great importance to building an ecological civilization and has carried out a series of fundamental, pioneering and long–term work for ecology and environmental protection as well as green and low–carbon development. He concluded the recent achievement of CRAES in scientific innovation and decision support from three aspects, including serving the top–level design of ecological and environmental protection, facilitating a comprehensive green transformation of social and economic development, and supporting "in–depth efforts" to cut pollution. He suggested further strengthen cooperation based on issues of common concerns, as promoting academic exchanges and sharing of countermeasures in the field of climate change synergizing the reduction of pollution and carbon emissions; maintaining extensive cooperation around the improvement of regional environmental quality including regional air pollution control, water ecology and environmental health; giving full play to the role of science and technology innovation support to contribute to solving real–world problems and enhance the exchanges of young researchers.

President KIMOTO Masahide expressed his gratitude to President LI Haisheng and his delegations for organizing TPM19 online, and President KIM Dong Jin and his delegations for their active participation. He noted the 50th anniversary of diplomatic relationship between China and Japan this year and spoke on various research activities. During these years, NIES has collaborated with CRAES on a wide range of research projects under these agreements and has set good examples of how researchers and government officials work together under international collaboration on various environmental issues. NIER has also been our joint research partner, particularly the study on Long-range Transboundary Air Pollutants in Northeast Asia led by NIER among the three countries, as well as research on Environmental Health field to assess health impact of the environment. Under NIES' Fifth Five-Year Plan, we aim to promote social implementations including policy contribution. NIES has participated in COP 27 to raise awareness and attract interests in decarbonization and adaptation to high officials around the world. President KIMOTO expressed his philosophy that there exist no boundaries between countries on environmental issues and also acknowledged the importance of TPM and our roles as leading institutes in the Northeast Asia. He concluded his speech wishing to hear more in today's presentations and discussions on how TPM can contribute to the future.

President KIM Dong Jin made an emphasis on international cooperation, a key addressing environmental issues intertwined over the world, with marking 30 years of diplomatic relations between Korea and China and 50 years of diplomatic relations between China and Japan. He has mentioned that global research cooperation becomes wider in the atmospheric field and that he is proud of fruitful results through NIER and CRAES' Joint Research Group aimed at reducing air pollutants and implementing carbon neutrality. Against this backdrop, he has said that the three national research institutes of NIER, CRAES and NIES should strengthen their global cooperative partnership in what is deemed a key to addressing issues, with sharing NIER's three strategic focus areas. President KIM has remarked that NIER's first focus is to widen future-oriented research in climate change and going toward carbon-free society by building bases to observe greenhouse gas emissions at a national level with the foundation of the National Adaptation Center for Climate Crisis (NACCC). He added the second focus of creating a pleasant environment for people's better life by putting energy into developing convergence of research with the fourth industrial technology. Last but not least, he has highlighted the environmental safety net for the public health, adding NIER has made concerted efforts to come up with measures to relieve stress from noise and light pollution for people's health.

The Presidents recalled the history of tripartite cooperation among China, Japan, and Korea, as well as the bilateral collaboration between both sides, to commemorate the 50th anniversary of the normalization of diplomatic relationships between China and Japan, and the 30th anniversary of the establishment of diplomatic relations between China and Korea. It was commonly recognized that the establishment of TPM was inseparable from sound bilateral cooperation, and TPM has been indeed playing a positive and effective role in promoting the sustainable development of Northeast Asia, since its establishment in 2004.

The Presidents listened to the results of the "Comparative Study on Climate Change", in terms of observation by CRAES, mitigation and public awareness by NIES, as well as impact and adaptation by NIER. They agreed that the current research results of the three institutes could be used for in–depth research cooperation on climate change, in particular, to exchange and discuss how to synergize the reduction of air pollutants and carbon emissions. They noted that the "Comparative Study" among the three institutes was an effective way of exploring the common interest and seeking further concrete cooperation under TPM.

The Presidents approved the changing mechanism of Leading Institutes (LIs) for the Potential Research Areas (PRAs), in which the LIs will be changed every three years for activating new vitality in the PRAs. In the new term herein, the LI of Air Pollution will be changed from NIER to CRAES, the LI of Environmental Health will be changed from NIER to NIES, while that of Water Environment be changed from CRAES to NIER, and the LI of Climate Change remains to be NIES.

The Presidents discussed future cooperation based on the progress reports and plans delivered by new LIs of PRAs. They agreed that synergizing the reduction of pollution and carbon emissions, and coordinated control of $PM_{2.5}$ and O_3 could be future cooperation focuses in the field of air pollution, while more advanced technologies and practices could be shared among the three institutes in the field of water. In terms of climate change field, the Presidents decided that research centers for climate change in the three institutes could enhance communication and cooperation concerning the observation of carbon and modeling for climate change assessment, as well as adaptation. As for the environmental health field, they pointed out that there are broad prospects for cooperation among three institutes, which could be further explored. Overall, the Presidents proposed that progress and achievements of TPMs could be reviewed and published or open to the public at a wider level.

To discuss the common concerns of Northeast Asia, the Parallel Workshop (PWS) for TPM19 was held focusing on "Atmospheric Observation and Source Apportionment based on Emerging Technologies". In the PWS opening speech made by President LI Haisheng, the achievements and effective approaches for air pollution control in China guided by the philosophy of ecological civilization were presented. President KIMOTO and President KIM recognized the achievements of China in air pollution control. The Presidents emphasized that concrete cooperation could be further strengthened among the three institutes for that air pollution is not an issue for a country but for the region.

President KIMOTO offered to host TPM20 in Japan. The specific date and format for TPM20 will be discussed and confirmed by the Working Group. The Presidents entrust Working Group to have detailed discussion in the coming year.

Finally, the Presidents expressed their appreciation of the fruitful outcomes of the meeting. President KIMOTO and President KIM extended their sincere gratitude to President LI and his staff members for their successful organization of the TPM19.

附录

本书缩略语中英文对照表

英文缩写	英文全称	中文含义
TPM	Tripartite Presidents Meeting among CRAES, NIES and NIER	中日韩三国环境科学研究院（所）长会议
CRAES	Chinese Research Academy of Environmental Sciences	中国环境科学研究院
NIES	National Institute for Environmental Studies, Japan	日本国立环境研究所
NIER	National Institute of Environmental Research	韩国国立环境科学院
TEMM	Tripartite Environment Ministers Meeting	中日韩环境部长会议
PRAs	Priority Research Areas/Potential Research Areas	优先研究领域／潜在研究领域
FP	Focal Point	联络人
LI	Leading Institution	牵头协调机构
LTP	Long-range Trans-boundary Air Pollutants in North East Asia Project	东北亚大气污染物长距离跨界输送项目
SLCPs	Short-Lived Climate Pollutants	短寿命气候污染物